The Flight of the Snow Geese

By the same authors

Nature's Paradise – Africa
Growing Up With Animals

The Flight of
the Snow Geese

Des and Jen Bartlett

COLLINS and HARVILL PRESS, Toronto and London, 1975

ISBN 0 00 262233 5

Set in Spectrum

Made and printed in Great Britain by
William Collins Sons & Co Ltd, Glasgow,
for Collins, St James's Place and
Harvill Press, 30A Pavilion Road,
London SW1

To our many friends at Survival Anglia Limited who are dedicated to making honest nature films for family entertainment and enlightenment.

Contents

Illustrations

* *photo by Les Bartlett*

BLACK AND WHITE

* photos by Les Bartlett
† photo by Julie Bartlett

Foreword

I first met Des and Jen Bartlett when they were working for Armand Denis in Africa. They impressed me as having a rare combination of gifts; they were devoted and skilled naturalists, and outstanding photographers who had a tremendous interest in all living creatures. Most important they were and still are – after all their great success – modest, friendly and delightful people.

This book describes their observations of the remarkable annual migration of the snow geese over the 2,500 miles that separate the bleak Canadian tundra from the Gulf of Mexico. It also tells of the challenge which taxed their powers of endurance and ingenuity to the full. Their account of these mysterious and spectacular birds adds greatly to our knowledge of snow geese and of the men who help to conserve them.

PETER SCOTT

Slimbridge
May 1975

From Siberian
breeding grounds

GREENLAND
(DENMARK)

ALASKA

ARCTIC CIRCLE

Baffin Island

CANADA

Mc Connell River Camp

Southampton
Island

Hudson
Bay

Churchill
La Pérouse
Bay Camp

James
Bay

The Pas

PACIFIC FLYWAY

Pembina

Sand Lake

De Soto

Squaw Creek

UNITED

CENTRAL FLYWAY

STATES

MISSISSIPPI FLYWAY

Tucson

Houston

Gulf of
Mexico

CUBA

MEXICO

**Autumn
migration routes**

Major breeding grounds

Major wintering grounds

National Wildlife Refuge

0   1000

Statute miles

© 1973 National Geographic Society
Originally drawn by Alfred L. Zebarth
compiled by Harold A. Hanson

Introduction

For many people, any mention of snow geese immediately brings to mind Paul Gallico's moving story of a lone snow goose blown by a gale across the North Atlantic to England. In Europe one of these birds is a rarity, but in North America the lesser snows, *Chen caerulescens caerulescens*, are the most abundant species of wild goose, numbering as many as two, or perhaps even three million birds.

Although lesser snow geese are so numerous, most people living in the United States and Canada have never seen one; this is because these birds concentrate in vast flocks which use the same flight paths, year after year, between the Arctic breeding grounds and their wintering spots along the Gulf of Mexico, well away from areas of dense human habitation. In fact, a less numerous sub-species, the greater snow goose, *Chen caerulescens atlantica*, is probably more familiar. Numbering little more than 150,000, these birds are larger than the lesser snow goose, and their flight path, the Atlantic Flyway, passes over the more densely populated areas of Canada and the United States. However, thanks to modern science, the lesser snow goose has now entered the homes of more than twenty million American families with the two television showings of *The Incredible Flight of the Snow Geese*, which Glen Campbell narrated and for which he sang the ballad 'Fly High and Free'. Elsewhere around the world, the same Survival Anglia film was shown with a commentary by Sir Peter Scott, the world authority on waterfowl. Television shrinks the world, and it is hard for us to visualize the 100,000,000 people who have marvelled at the close family lives of the snow geese, and have thrilled as we have done to the sight and sound of thousands of geese tumbling out of the sky to the safety of a wildlife refuge.

Two years of hard work – or more aptly, two years of fascinating endeavour – went into making the TV Special, but in a one-hour film it was impossible to tell the full story of the snow geese or of the many other creatures which share their world. In this book we shall try to tell the story of our lives with the wild geese, our adopted snow goose 'children' and Fred, the sandhill crane with personality plus.

The romantic snow geese long remained a mystery to scientists and laymen alike. Each spring, as soon as the thaw begins, vast formations of snow geese – often accompanied by blue geese, a colour phase of the lesser snow – pass over the United States and Canada to disappear in the Far North for four or five months. The original scientific name for the lesser snow goose – *Anser hyperborea* – indicated that their breeding grounds were somewhere 'beyond the north wind'.

Snow geese are well named, not only because of their white plumage, but because snow governs the very rhythm of their lives. In the autumn they head south for the winter months, and reappear over settled areas before the first heavy snowstorms whiten the ground. As the geese fly past, a thousand feet or more above the ground, the hauntingly wild sound of their calls stirs the imagination of many a youngster lying in bed at night on the lonely prairie farmlands.

Our own introduction to snow geese came several years ago during a visit to wildlife refuges in the Klamath Basin, along the border between northern California and southern Oregon. We had always hoped to see this area one day, where North America's most spectacular concentrations of waterfowl are to be found in the autumn. Originally the Klamath Basin, or valley, contained large shallow lakes and marshes, attracting the great flocks of ducks and geese migrating each year along the Pacific Flyway (or route). Here they paused to feed and rest before continuing southwards. During this century most of these wetlands have been drained for agricultural purposes, and although this has not altered the traditional migration routes, the waterfowl now have to crowd into the remaining suitable habitat, mostly on wildlife refuges.

Unfortunately, our arrival at Tule Lake Refuge, headquarters for the complex of federal wildlife refuges in the area, was delayed until late November. We arrived to find a few inches of snow on the ground and much of Tule Lake frozen over, although ducks, swans and geese still swam in the remaining narrow channels of open water.

In October there had been five million waterfowl in the area, but we were told there were now only half a million left, as most of the birds had moved on south early in November. In almost any other part of the world, half a million birds would be considered a unique spectacle. We were thrilled, for it excelled anything we had experienced with waterfowl in Australia, Africa, or North America up to then. Ducks seemed to be the most numerous and of the geese the small species of Canadas, known as cackling geese, far outnumbered the snows.

Next we moved south to find the main concentrations of ducks and geese on a group of refuges in the Sacramento Valley of California, where the most abundant species of geese were the snows. Here the Sacramento Valley is very flat and open, distantly flanked on the east by snow-capped mountains.

Plagued by valley fog, we were lucky to have one fine day a week for filming. We never ceased to be impressed by the sight of several thousand snow geese resting on open ground together, or feeding in a watery rice field. When alarmed, they all took to the air simultaneously with a roar of wings and a clamour of voices – an overwhelming combination of sight and sound. Almost as spectacular was the scene of thousands of ducks rising from a pond in a huge dark cloud, only to wheel in a wide circle and settle on the water again.

Fascinated by the gregarious snow geese, with their gleaming white bodies and black-tipped wings, we began learning all we could about them, with the idea of one day making a detailed film of their life cycle. We found that the snow geese using the Pacific Flyway on their autumn and spring migrations nest in colonies along the Arctic shores and islands of western Canada and eastern Siberia, but hardly at all in Alaska. Any hopes we had of following them to their summer nesting grounds were dashed when we discovered that the main snow goose colony was close to the site of a Russian rocket range!

For the next few years, while other filming projects kept us busy in Canada, the United States, Mexico and South America, we continued to gather information about the snow geese. Dr Graham Cooch of the Canadian Wildlife Service, a renowned snow goose expert, was very helpful and gave us much valuable advice. Unlike most species of water-fowl, snow geese are colonized breeders, and we learned that there are seven major nesting colonies around the shores of Hudson Bay. The most accessible of them, Dr Cooch told us, was at La Perouse Bay on the south-west shore of Hudson Bay, only thirty miles east of Churchill, but if we wanted to film geese in a typical tundra setting, he advised us to go to the McConnell River, almost 200 miles north of Churchill.

Our plans gradually began to take shape. We learned that Dr Charlie MacInnes, an Associate Professor of Zoology at the University of Western Ontario, took a small group of students to the mouth of the McConnell River each summer to study the nesting geese and other wildlife. He was enthusiastic about our idea and helped us immensely with advance planning, and later with valuable 'on location' advice. So began what one magazine writer has called 'The World's Longest Wild Goose Chase'.

Food cache
on shore
of north
branch of
the river

Ross's
Goose nest

Golden Plover ●

Camp
Esker

Crane raiding
Snow Goose
nest

Tower
blind
for
Canada
Goose study

Crane nest ●

Mixed ● ●
pair

Snow Goose ■
Blind

Tower blind ··· ■

Improvised
landing
strip CAMP ▲

● Polar Bear

Arctic Loon's nest

Red-throated
Loon's nest

● King Eiders

Jaeger's nest ●

HUDSON

BAY

Mouth of the Mc Connell River
North West Territories Canada 60°50'N 90°25'W

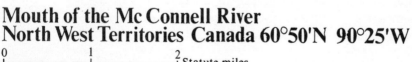

0   1   2
Statute miles

>>>>>> Esker

1 From Desert Heat to Arctic Cold

As we expected to be cut off from the outside world for almost four months while camped at the McConnell River during the northern summer, equipment for the expedition had to be thoroughly checked for there would be no running to the neighbourhood hardware store for forgotten items. After filming in Mexico, we paused in Arizona to re-equip for colder climates. In the ninety-degree desert heat of early May, it wasn't easy to think of packing such things as ice scrapers for the car windscreens, down-filled clothing and sleeping bags. But within a week of leaving Arizona we were to appreciate such items. Nothing gives one a better idea of the vast size of North America and its geographic range than a road trip north at this time of year. After eight days we reached the McConnell River. We had experienced a temperature range of 91°F to 6°F.

Knowing that we would need help on the tundra in order to accomplish as much filming as possible during the short northern summer, we arranged for our eighteen-year-old nephew, Les Bartlett, to fly over from Australia. He needed little urging to take leave of absence from his nine-to-five bank job to join us, and it is doubtful if he will ever again be able to settle down to an office routine. Our little group was complete when our friend Lee Lyon from Palo Alto, California, joined us. She had already been with us for six months during the Sea of Cortez filming in Mexico and was therefore almost a member of the family – an important consideration since we would be cut off from other people for so long. Lee was to help with some behind-the-scenes filming and with raising our family of young goslings.

Finally, on 8 May, we left the desert warmth, where prickly pear cactus and other desert wild flowers were blooming and headed north-wards in two heavily laden vehicles – a Chevy station wagon, and a 1959 Land-Rover we had used in Africa. We enjoyed telling people 'The Land-Rover is almost run in . . . it's done just 182,000 miles!' Our journey by road took us through the centre of the United States from south to north, then into the Canadian province of Manitoba, where our first destination was The Pas, two and a half thousand road miles from Tucson, Arizona. This marked the end of the paved road and was the most convenient place for boarding the train that would take us a further five hundred miles to the outpost of Churchill, on Hudson Bay.

The drive north was almost relaxing, for there was no need to rush. Or so we thought, until we paused early one morning in South Dakota to telephone Charlie MacInnes in Ontario.

'I'm very glad you called,' said Charlie. 'Some of my students have gone on ahead to Churchill and I've just heard from them that there's been an unusually early thaw up there. Unless you can reach The Pas by noon tomorrow, you'll miss getting your vehicles and gear on the weekly freight train. Then you might have real difficulty reaching the McConnell River this summer. The only way to get there at this time of year is to land before the snow thaws, in a plane equipped with skis.'

This news came as a shock, but I figured that it was just possible to catch the freight train if we drove day and night, without stopping to sleep. Charlie was to fly later that same day to Churchill, and would be there to meet the train when we arrived. But if we were to reach The Pas by noon next day, we could not afford a breakdown or too many punctures!

Considering the length of the trip our timing was perfect – we reached The Pas with half an hour to spare and heaved a sigh of relief as the heavy sliding doors of the box cars were closed and sealed, with our two station wagons safely inside.

It is doubtful whether any other four passengers boarding the 'Muskeg Express' at The Pas next morning enjoyed the trip more than we did; it was sheer bliss to sit back in the comfortable train seats and leave the driving to others during the twenty-four hour trip to Churchill. At first the railway track passed through thick stands of spruce, aspen and birch, and we noticed many places where beavers had been at work felling large deciduous trees. The farther north we went, the smaller the trees became, with here and there wide areas of stunted conifers. Although our life is one of almost constant travel, we were all as excited as school children on a picnic as the train neared Churchill next morning and we had our first glimpse of the tundra. Now there were patches of thawing snow, with just a few stunted spruce and tamarack standing here and there on open sections of tundra type vegetation. The ground-hugging plants were tinted various shades of brown, red and yellow, with scattered pools of melt-water reflecting the blue of the sky or the white of scattered clouds. Our excitement quickened as we caught sight of our first willow ptarmigan, still wearing their white winter feathers. Our northern adventure had really begun.

True to his word, Charlie MacInnes met us in Churchill and enabled us to make more arrangements in a few hours than we could have done in two days on our own. It was his tenth summer in the Arctic, and he seemed to be friends with everyone in this northern outpost. However, it was still touch and go whether we would be able to fly in to the McConnell River, owing to the exceptionally early thaw.

Charlie's small group of students had already flown with all their supplies on the regular D.C.3 plane from Churchill to the small settlement of Eskimo Point on the west shore of Hudson Bay, 200 miles to the north and only 30 miles from our eventual destination. Three of the students and their Eskimo guides then pressed on to the mouth of the McConnell River with small snowmobiles, in order to mark out a landing strip for the ski-plane on the best snow they could find close to camp. We heard later that they had a hairy trip over the sea ice of Hudson Bay, travelling at night – it never gets completely dark at this time of year – because then the ice would be at its firmest. Even so, one snowmobile broke through the sea ice, upsetting the *komatik* (sled) that it was pulling and giving the two Eskimo passengers a thorough wetting in the icy water. Their clothes were stripped off instantly in below freezing air, and the students' dry garments handed out to the unfortunate pair. It seemed curious to us that the Eskimos, whose lives are often endangered in the harsh climate of their homeland, should be travelling at this time of year without spare clothing of their own.

We had arrived at Churchill on Saturday 15 May, just one week after leaving sunny Arizona. We immediately needed our down jackets to protect us from the icy winds blowing off the frozen expanse of Hudson Bay. The following morning our vehicles and equipment arrived on the freight train. We drove to the airport and began weighing our many items of equipment in a Lambair hangar, as we needed to know how many plane loads were involved. Because of the uncertainty of the snow conditions for landing at the McConnell River camp, the pilot planned to carry less than a maximum load on the first trip. Our total weight, including people and equipment, was under 2,500 pounds, much less than we had anticipated and no problem for the single-engined Otter aircraft to carry in two trips. We took very little food, as a few weeks earlier we had accepted Charlie's offer to increase his bulk food order to cover our needs.

We could hardly contain our excitement as the chartered Lambair Otter took off from Churchill on 16 May, with ourselves and the first load of carefully selected gear aboard. As the plane banked to head north we had a clear view of the wide expanse of the frozen Churchill River, its uneven, jumbled surface glistening with shades of blue and green wherever the frigid winter winds had blown the snow cover clear of the river ice.

I had the best view of all as we flew steadily north, sitting in the co-pilot's seat so that I could slide the window back for filming. Surrounded by piles of equipment in the cramped main cabin of the plane, Jen, Lee, Les and Charlie all eagerly watched the passing scene below. We flew above the western shore of Hudson Bay, but much of the actual shore-line was indistinguishable, as the snow on land had the same appearance as the snow-covered sea ice. Here and there snow-free patches of bay ice formed beautiful patterns in various shades of blue, with sharp ridges and dips where the first thin ice of early winter had been broken by wave action and refrozen solid to a depth of several feet. This solid sea ice does not break up completely until mid-July, and some years even later. Before long there was a shout from Charlie.

'There they are,' he called out. 'Our first snow geese flying north.'

Looking down we could only see the dappled brown and white landscape of the thawing tundra.

'Wait a moment and they'll pass over that brown patch of tundra,' Charlie shouted above the roar of the engine.

Far below, long lines of white specks flew over a brown area and also across dark patches of open blue water where the Seal River had begun to thaw near the coast. As soon as the geese flew above patches of snow, they became invisible to us. The plane droned on and we noticed that the trees decreased as the snow increased. After flying for about an hour, Charlie called out again.

Take a good look to the west. Those are the last trees you'll be seeing for almost four months!'

Now we were really in the Canadian North, having crossed from Manitoba into the Keewatin District of the North-West Territories. We could no longer see geese flying below, for we had overtaken their northward migration. But if the thaw kept up the birds would not be many days behind us.

After we had been airborne for almost two hours, Charlie shouted 'There it is!'

Peering down we could see nothing but the bleak and featureless tundra, almost completely covered with snow. Then the plane began to lose altitude and circle. We soon made out the line of the partly thawed McConnell River, and finally the dark pinpoint on the snow that was the one-room plywood hut built by the biologists two years earlier. Nearby, the advance party of students had used orange marker flags to indicate the most suitable landing strip on the snow. The pilot circled a few times to make sure it was safe to land, then made his final approach. A bumpy landing on the ridgy, crusted snow and we were there. Our home for the next three and a half months would be a draughty tent on the windswept tundra.

We stepped from the plane into a dazzling world of white and wondered momentarily what on earth we had let ourselves in for, coming to such a desolate place. Luckily the sun was shining, and although a cutting, icy wind was blowing, our first impressions of the area were formed under near optimum weather conditions. However, there was little time for reflection; Charlie introduced us to the students and we all set to work unloading the plane.

For the first two nights we all slept in the biologists' cabin, for our chosen tent site was just off the end of the short snow runway and it was too risky to erect the tents until the pilot had brought in the final load. Once he had taken off for the last time, our major task was to set up our two tents and carefully stack all the equipment inside. Fortunately, about seventy yards from the biologists' hut there was an area almost free of snow where we could erect the tents on the bank of the McConnell River. The camp was on a very low stony ridge, only a few feet above the surrounding area, but high enough to remain fairly dry when the rest of the thawing tundra became a watery maze of small lakes and boggy areas. Because of the permafrost* the ground was rock-hard and

* Permanently frozen ground

tent pegs were useless. To anchor the tents securely against the almost constant wind, all guy ropes had to be tied to piles of large rocks. Even these were frozen to the ground and had to be broken free before we could move them into position. Light snow was falling as we put up our main tent, which had been made in Kenya some years earlier. Although originally designed for warmer climates, it surprised everyone by surviving the summer almost unscarred, in spite of violent storms with winds gusting to forty or fifty miles an hour. Because of condensation problems when the air temperature warmed a little inside the tent, all supplies had to be kept up off the cold, waterproofed floor. Also, to keep fine spray off during the most severe rainstorms, everything inside the tents had to be covered with plastic sheeting.

For our first few weeks at the McConnell River the evening temperatures were sometimes only a few degrees above 0°F with the daytime reading often failing to rise above freezing, 32°F. To clean our teeth, or have an early morning wash in cold weather, we first had to break the ice at the edge of the river, for although the thaw had begun on our arrival, the shallow edges refroze most nights. Because of the extra wind-chill factor, a partly filled glass of water placed on a nearby boulder would be frozen to the rock by the time it was needed for a mouthwash, and the bristles of a toothbrush froze solid only seconds after the cleaning job was done. Little wonder that I gave up shaving on the tundra and have worn a beard ever since! Our evening warm water sponge before retiring was a hasty affair in the unheated tents, but we soon warmed up in down sleeping bags. However, none of us raised any objections to Charlie's suggestion that we should eat most of our meals with his group in the comparative comfort of the hut, which was warmed by an oil-burning heater. Here we spent many pleasant hours, especially during bad weather, learning a great deal about tundra life from Charlie's vast store of experience and from text books. His team consisted of four graduate students, Dave Ankney, Allan Aubin, John Harwood, Connie von Barloewen and three assistants, Dave's wife Jeannette, Larry Patterson and Jim Izawa – none of them had any previous experience in Arctic regions. Dave and Jeannette had their two-year-old daughter Theresa staying for the summer, and this kept Jeannette at base, where she cooked the evening meal for everyone. We certainly appreciated this luxury after being away filming for up to fifteen or sixteen hours on bright days.

Willow ptarmigan were our first visitors, wandering among the guy ropes of our tents and showing no fear of humans. In May most of the female ptarmigan were still in their white winter plumage, although on a few the brownish summer feathers were just beginning to show. The males had white bodies and wings, with the head and neck a deep reddish-brown. We were fascinated by the way the white feathers entirely covered their legs and feet to give them greater insulation. Unlike most other species of birds that use the tundra for summer nesting, these ptarmigan do not migrate far to the south in winter. Somehow they manage to survive all the year round in these latitudes, digging in the snow in winter to feed on tiny willow buds and sheltering in snow burrows from the icy winds. We watched one female ptarmigan lying down in a natural hollow in the snow to have a 'snow bath'. Loosening the snow with her feet before wriggling down into it with feathers ruffled and wings partly open, she flapped gently, and appeared to enjoy herself immensely. A few weeks later, when the low camp ridge became dry enough for the ptarmigan to dustbath, we saw this same action daily, in dust instead of snow. After loosening the sandy soil, a ptarmigan would roll slightly on to its side, then use one foot to kick dirt up on to its feathers. Soon the area near our tents was littered with these bowl-shaped depressions. In the early morning hours we often awakened to the loud cackling calls of a male ptarmigan as it flew on to the ridge pole of our tent, the highest point around. From this vantage point he noisily proclaimed his superiority. Usually his performance was interrupted by a second male flying up to challenge him, and we drowsily watched their shadows as the birds slid down the icy canvas.

During our first days in camp we saw a few small groups of sandhill cranes, Canada geese, whistling swans, and snow and blue geese calling musically to each other as they flew northwards. It wasn't until we had been on the tundra for a full week that the geese began coming in large numbers. 23 May began as a calm, foggy morning, with long crystals of hoar frost hanging from our tents. By 10 a.m. the sun was out and the biologists estimated there were 300 geese flying past the camp every hour. Some of these were Canada and white-fronted geese, but the majority were snow and blue geese, the snows outnumbering the blues by seven to one. At the peak of the spring migration the number of blue geese flying over increases until there is one for every three snows, with up to 1,000 geese passing the McConnell River in an hour.

For many years scientists classified the lesser snow goose and the blue goose as two distinct species, but after detailed studies of the birds on their breeding grounds and in captivity, it has been proved that they are merely colour variations of the same species. The blue goose was the first of the two to be scientifically named, by Carolus Linnaeus in 1758, when he called it *Anas caerulescens*; *caerulescens* means 'bluish' in Latin, and has now become accepted as the specific name for both colour phases of the lesser snow goose. Scientists are still divided about the most appropriate generic name: *Anser*, if one believes the snows and blues to be part of a group of typical 'grey' geese; or *Chen*, if regarded as being sufficiently different to warrant special distinction. At the moment *Chen* has become accepted, although many leading waterfowl scientists still prefer *Anser*.

The largest nesting concentration of these geese – over a million birds – gathers each year on the Great Plain of Koukdjuak on the west coast of Canada's huge Baffin Island. This multitude, consisting largely of blue geese, migrates along the eastern side of Hudson Bay, hundreds of miles to the east of the McConnell River, (see map page 16). Most of the snow and blue geese flying past our camp continued northwards, many of them bound for nesting grounds on Southampton Island. But gradually others began landing on their traditional nesting areas across the river from our tents. They were home for the summer, after a two-and-a-half thousand mile flight from their wintering grounds near the Gulf of Mexico. There is a great urgency in each spring's northern migration. The Arctic summer is so short that the birds must arrive and begin nesting the moment the tundra is sufficiently free of snow, otherwise their young may not be ready to fly south before the early winter storms are upon them. The nesting pairs of geese have just over three months in which to transform newly-laid eggs into self-sufficient airborne juveniles.

Why do these vast numbers of geese and other birds fly so far north to nest every year? On the vast stretches of tundra in the Arctic and sub-arctic, the birds can of course nest with little or no disturbance from man, and there are relatively few other predators to destroy their nests and young. But perhaps the main reason why so many birds use these remote nesting grounds is because of the abundant supply of food available at the time the young hatch. For although the summers are short in these northern latitudes, the growth rate of plant and small animal life is greatly accelerated for two or three months by the fact that, even at the McConnell River on clear days in midsummer, it never became completely dark; indeed there were sometimes almost 20 hours of sunlight.

Geese pair for life and mating takes place on the journey north, so that egg-laying can begin soon after they reach the nesting area. Two-year-old birds, breeding for the first time, choose a mate on the southern wintering grounds where birds from the different nesting colonies intermingle. For many years the theory was held that all geese returned to the nesting colony of their origin to breed. However, recent research by Dr Fred Cooke and his associates from Queen's University in Ontario, has shown that males rarely return to their colony, and will often follow the female to her home colony to nest. So maybe geese are the original leaders of the Women's Liberation Movement!

2 *The McConnell River*

Already the thaw had begun, and water was flowing in the river beside our camp. Vast ice bridges still covered many stretches of river, so that in places the water flowed out of sight; in others, where the rocky bed was still frozen, the river flowed over the ice, which appeared yellow from above due to a staining in the water. Throughout the day and evening we heard loud bangs and cracks, not unlike rifle shots, as huge chunks of ice broke away and fell into the water below. These massive pieces of ice were far too heavy to be washed downstream, and as the river flowed round them they were gradually eroded into spectacular shapes.

Parts of the bank of the McConnell River were at this time covered by ice cliffs up to five feet high, and on sunny days tiny rivulets of water ran over the edges to form miniature waterfalls. On clear nights this run-off froze into crystal stalactites and stalagmites, glowing like rows of golden jewels in the dawn sunlight, transforming our white world into a fairyland of sparkling colour, with not the slightest trace of haze in the clear blue sky. And there was nothing to interrupt the 360 degree panorama of this brilliant world: no trees, no mountains or even hills blocked our view of the flat white terrain, broken only here and there with patches of brown thawing tundra. A few rocks showed up across the tundra, but there was absolutely no way of knowing how big they were, for size is relative to distance, and there was no convenient tree or building on which to base scale. No doubt it was an even more dramatic experience for the astronauts on the surface of the moon.

Here on the tundra we had the feeling that the world was indeed round, as one could slowly turn and see the flat horizon stretching unbroken for a full circle. In this northern area an esker is the only thing to break the monotony of the scene. Having grown up in Australia, it was not surprising that we had never heard of an esker before. Charlie was again our tutor.

'During the last great Ice Age,' he said, 'this whole Hudson Bay area was covered with ice two miles thick.'

As the ice began to melt, creeks and rivers formed, flowing on top of the ice itself. Sand and stones trapped in the ice were washed into the rivers, forming a growing layer beneath the flowing water in certain places. When all the ice finally melted the accumulation of stones and gravel was deposited on the otherwise flat ground, forming a low ridge, or esker. Camp Esker could be clearly seen almost two miles away across the McConnell River and it was the first esker we were able to visit.

Months earlier, when I first went to see Charlie MacInnes at the University of Western Ontario, we had discussed all terrain vehicles for moving heavy supplies across snow and ice. We decided it would be a big help all round if I bought the Canadian-made Argo. This was sent by train to Churchill, then by a large commercial plane to Eskimo Point, and finally driven down to the McConnell River by three of the scientists, travelling mostly over the sea ice.

Charlie had made prior arrangements with Eskimos to bring in some of the heavy camp supplies from Eskimo Point by snowmobile in the early spring. These included drums of gasoline and large cylinders of bottled gas. All went well until the early thaw prevented the snowmobiles from crossing the north branch of the McConnell River with the final loads, which had to be cached on the bank. While snow and ice still covered most of the ground, we offered to ferry these supplies the three or four miles back to camp. Half a mile upstream from camp there was enough open water to make a crossing of the main river in the Argo, using an outboard motor on the back. Once the wheels touched bottom on the far side of the river the outboard motor was lifted clear of the water and the sturdy little machine climbed the bank over ice, rocks and boggy ground, its eight balloon tyres filled with an air pressure of only two pounds per square inch.

En route to the supply cache we were able to visit Camp Esker. Leaving the Argo at the base, we walked up the sixty-foot high esker and met two Eskimos at the top. One was dressed in the traditional caribou skin clothing, while the other wore a western style parka and spoke a few words of English. They were searching for caribou to hunt and asked to borrow our binoculars, but this wasn't their lucky day. They had camped the previous night near the base of the esker, simply turning the *komatik*, which they had pulled by a snowmobile, on its side and stretching a piece of canvas out from the lee side before crawling underneath to sleep. We noticed the remains of an Arctic hare at their camp site and during the following weeks we were to see its lonely white mate many times. Conspicuous on rocky ground, this large white hare became almost invisible when moving across the snowbank that clung to the north slope of the esker for most of the summer. If we approached very slowly in order to photograph it, the hare allowed us to come within ten yards.

Moving the supplies back to camp proved hard work, and took several trips in the Argo. The north branch of the McConnell was not deep, and wearing hip waders we pushed the Argo across. Much of the crossing had to be made walking on a layer of ice at the bottom of the river and this was a very slow and slippery business. Once we had made one safe crossing we tried to use the same place on subsequent trips. With the Argo fully loaded with supplies, there was no room for passengers. I wanted everyone to take a turn at driving, but the others elected to walk so that I would be the one to get the Argo stuck if this should happen. Amazingly enough, even with such heavy loads, we were always able to coax it across the difficult terrain, but sometimes on very slick surfaces the others had to push.

Our last trip for supplies in the Argo coincided with the break-up of the remaining ice in the McConnell River, and a prettier sight would be hard to imagine. Most of the floating ice had become rotten and 'candled', forming vertical candle-shaped pieces, often a foot or more long, which tinkled musically and broke loose whenever two ice cakes collided. Other formations floated upside down, their exposed undersides displaying gleaming pinnacles of ice that resembled sparkling chandeliers as they sailed majestically by. More solid blocks of ice rolled end over end downstream in the section of rapids. The heavily loaded Argo looked like a floating bathtub, with only six inches of freeboard above the icy water, as we gingerly made our way home across the river. We shall always remember this hair-raising trip in late May, for the element of danger as much as for the sheer beauty of those alarmingly large chunks of gleaming ice floating and tumbling past us in the swift current.

Following the final break-up of ice in the McConnell River we at last felt that spring had really arrived. Snow and ice would still be around for many weeks, but we knew this should become imperceptibly less with each passing day. We had mounted a movie camera on the side of the Argo to film travelling scenes, and one day I asked Charlie to drive across a frozen lake while Les and I sat in the back to operate the cameras. A sudden application of the brakes to the wheels on only one side of the vehicle while it sped across the ice would make the Argo spin crazily on the glassy surface and this was great fun, giving everyone on board a thrill. This time, however, the ice was thinner than we had anticipated and in mid-spin the Argo suddenly broke through, coming to an abrupt stop while travelling sideways and throwing Les and me through the ice as well. Luckily the lake was no more than waist deep and camp was only a few hundred yards away, so our freezing clothes were soon changed for dry ones. Later, we all had a good laugh about the incident, but had this taken place a long way from camp it could have been far more serious. We were fortunate that during the whole summer nobody had a bad accident or illness, for although we carried a well-stocked medical kit, such things as hospitals and doctors were far away.

Our drinking water came straight from the river beside camp and we weren't at all put off by the fact that it was stained a yellowish brown by the tundra plants. Charlie's students of earlier years had given it two nicknames: 'Lemmingade' and 'Goose Juice'. We were sorry that this was not a lemming year; we did not see one all summer. But we soon realized the truth of the 'Goose Juice' title. Soon after the geese arrived in force on the tundra their droppings littered the bottom of every pond. This was ultimately our drinking water, once the ponds had slowly drained into the McConnell River!

On the rapidly thawing tundra the gregarious snow geese stake out their territories, the pairs often fighting neighbours to defend their chosen areas. Experienced nesters from previous years are probably the first to arrive, and immediately settle down to nest-building while patches of snow still whiten the tundra around them. Marked geese have been known to return to exactly the same nesting area year after year, although a human would have difficulty locating any given spot within half a mile, as the many ponds forming on the flat tundra seem to change from week to week, let alone from year to year. The one-year-old geese, hatched the previous year, have also flown north with their parents, but are now separated from the nesting adults and band together on the outskirts of the colony near the coastal flats. This group of 'non-breeders' also includes any older geese which for one reason or another are not nesting. Nesting requires excellent physical condition, with ample reserves of fat to carry them through the incubation period without feeding.

Snow geese are large birds weighing between five and seven pounds. Surprising as it may seem with thousands of geese around, we found it very difficult to find an active goose nest on the open tundra during the early egg-laying period. Sometimes a goose would use an old nest from the previous year, laying its eggs in the central depression and covering them with layers of damp grasses and mosses. Old feathers and down are nowhere to be seen, but the circular nest is still slightly higher than the surrounding ground. A completely new nest is even more difficult to detect, for the vegetation covering the eggs appears no more disturbed than nearby areas where the geese have been feeding and digging in the spongy ground with their bills.

It was Lee who found the first active snow goose nest on 29 May. Poking around in a seemingly ancient and neglected nest she was surprised to discover a slightly muddied, once white egg buried beneath four inches of old, damp vegetation. It hardly seemed possible that this three-inch-long egg could be newly-laid, but after carefully covering it again, we placed one of our small orange marker flags several yards from the nest. The following afternoon we returned to check. As we approached slowly, a pair of snow geese retreated a short distance so we made our way to the nest and lifted the covering layer. Sure enough, there were now two eggs hidden in the nest. After carefully concealing the eggs once more, we erected a small tent as a blind forty feet to the south of the nest and quickly left the area so that the pair of geese would return to their territory.

On subsequent days we were able to take turns to watch from this blind as the goose returned to the nest to lay two more eggs. Spending little more than half an hour on the nest each time she laid an egg, our goose used her bill to build up the sides of her nest gradually with pieces of twigs, grass, moss and lichens, all plucked from within her reach immediately around the nest. We learned from the biologists that the geese normally lay one egg a day, but often miss a day before laying the final egg in their clutch. Not until our goose had laid her full clutch of four eggs did she begin to incubate, and only then did she line the nest with down from the underside of her body. Soon the nests were easy to see as bits of fluffy white down blew about and caught on nearby vegetation. Later we were to find that four or five eggs is the average clutch for the snow goose, although we did see one nest containing nine eggs, and three other nests with eight eggs each.

For a little over three weeks – 23 days to be precise – the goose kept her eggs warm day and night, while the gander stood guard nearby. If other geese came near he vigorously defended the nesting area and we witnessed and filmed many spectacular fights, with feathers flying, before the trespassers retreated. As territorial limits were established these fights became less frequent, and neighbouring geese in the colony settled down for the long period of incubation.

One day while our goose was on the nest laying an egg, we watched from the blind as a male ptarmigan landed a few yards from her. Immediately the gander flew in from twenty yards away and aggressively chased off the willow ptarmigan. Two weeks later, when incubation was well under way, this same gander took little notice when the male ptarmigan passed within a few feet of the sitting goose. Soon we discovered why we had so often seen the ptarmigan near this blind. Even though we had been visiting this area daily for over two weeks, it was not until Lee almost stepped on the amazingly camouflaged female ptarmigan that we found her nest within a few yards of our blind. On the open tundra no vegetation grew more than eighteen inches high, but the ptarmigan had carefully made her nest on the ground among six-inch-high birch twigs. By now she was in her summer plumage of speckled light brown, which blended in beautifully with her surroundings, making her almost invisible as she sat low on her clutch of eight speckled eggs. The male, by contrast, retained his now conspicuous white body and wings until later in the summer, presumably so that he could decoy any predators away from the sitting female. This was the ptarmigan we had already noticed using the top of our tent as a vantage point; he often flew there to make his loud cackling calls, even when we were inside. Once, unseen by the pair of geese, Jen was able to lean out of the door at the rear of the blind to take his portrait as he sat there, without in the least disturbing the vocal ptarmigan.

The nesting territory of 'our' particular pair of snow geese was about twenty yards in diameter, but we found others nearer the coast to be nesting within ten yards of each other. Nesting densities of up to 3,000 snow goose nests to a square mile have been recorded. Once her eggs were laid our goose remained on her nest except for brief trips to drink and bathe at small pools within her nesting area. Before leaving her nest she carefully covered the eggs with down and vegetation to keep them warm and hidden from the ever watchful gulls and jaegers* flying over-head. With so many geese concentrated on the nesting grounds, the available food in the area was quickly depleted, but until the eggs hatched neither goose nor gander left the nesting territory in search of food. During the period covering egg-laying and incubation the fasting geese have to rely on energy reserves built up during the preceding months of winter and spring when food was in plentiful supply. A nesting goose may lose twenty per cent or more of her body weight during this period, and in severe weather some females actually die on the nest, succumbing to a combination of freezing temperatures and starvation.

Several miles before it reaches Hudson Bay the McConnell River divides into two branches which then flow independently to the coast. Between the north and south branches of the river there is a large triangular island, three miles wide at its base on Hudson Bay, and extending inland for about seven miles. The main snow goose colony is spread over this island, with some geese nesting on the mainland both to the south and to the north. Our camp was on the southern bank of the river's south branch, some three miles inland from the coast, (see map page 16). Once the geese arrived in the spring there was a constant chorus of activity across the river from our camp. Day and night it went on, but when incubation began the noise gradually settled down to a steady murmur. Even this was easily heard from camp above the sound of the rushing waters of the swollen river.

After the geese had arrived at the McConnell our days settled into a gruelling routine. Each morning before 5 a.m. I looked outside the tent, and unless there was rain or dense fog I roused the others. After a quick breakfast, usually of pancakes or oatmeal, the four of us left camp by 6 a.m., carrying in our back-packs sandwiches and Thermos flasks as well as camera equipment. We often used the Argo in the early part of the season, strapping extra tripods and equipment on to the engine cowling.

Bread supplies taken to the McConnell River in May lasted until early July, keeping well in the outdoor deep freeze of a storage tent for the first weeks. Later, bread had to be baked every few days in the biologists' propane gas oven, with Jeannette and Charlie as the chief

* or skuas, as they are called in Britain.

bakers. Bought in bulk, Canadian freeze-dried foods were a boon, especially for evening meals, with excellent meats, vegetables and some fruit. Light and easy to carry, we found these far more practical and palatable than canned foods, although for variety some of the latter were included among the supplies. During our long stay on the tundra the foods we missed most were fresh fruit and eggs, but we were careful to supplement our diet with an adequate daily dose of vitamins. Later in the summer, Les and some of the students fished for grayling in the river and we all enjoyed these delicious fish, weighing between one and two pounds.

By mid-June at the mouth of the McConnell River, latitude 60° 50′N, longitude 90° 25′W, the sun sets for little more than four hours, with the pink afterglow of sunset merely moving along the horizon for perhaps 70° until the sun shows itself again around 3 a.m. For three months it never became dark, but we had no problem sleeping. The dark green canvas of our tent cut down the light considerably and anyway we became so tired that bright light alone would certainly not have kept us awake. Since filming was possible for up to eighteen hours a day if the weather was favourable, we often weren't back in camp until nine or ten o'clock in the evening. After eating a welcome hot meal with the biologists and washing the dishes, we struggled to stay awake long enough to complete the tedious task of sorting exposed film and writing up our notes, before crawling into our sleeping bags around midnight or later. Five o'clock came round all too quickly. We were almost sleep-walking after several weeks of this routine, but with so much to see and film during the short northern summer, we knew we had to make the most of the long hours of daylight.

By the end of May virtually all the geese had begun egg-laying, and during the first few days of June most started incubating. From a filming standpoint, this meant that all egg-laying and nest-building had to be filmed during this short period. Even more critical was the thought that we would have perhaps a week at the most to film the hatching of the goslings later in June. All we could do was hope for favourable weather for at least part of that period. Although Charlie had warned us to be prepared to make the most of every filming opportunity, the synchronization of these activities among 100,000 pairs of snow and blue geese really amazed us. Collectively, the whole colony completes egg-laying, incubation and hatching of the goslings in just over a month.

With nesting under way it wasn't long before we had erected eight blinds, or hides, within a three mile radius of camp. These little green umbrella tents were not easy to secure on the tundra where tent pegs were useless, even when the ground thawed, because of the soft, boggy conditions. To strengthen the blinds against the almost constant winds, we firmly taped all joins where the poles fitted together, and placed heavy rocks inside on the tent floor. Thanks to these precautions our blinds suffered little damage even during the worst storms. The green canvas of these small tents was visible for miles on the flat, treeless tundra, and made it much easier for us to keep our bearings, as some of the blinds remained in the same position for weeks. As mentioned earlier, it was hard to give scale to anything on the tundra: is that stone a huge boulder half a mile away, or merely a much smaller rock only three hundred yards away? On one occasion we thought we saw a caribou in the far distance, but on using binoculars, discovered it was a sandhill crane much closer to us. Even when we first arrived and there was so much snow on the ground, 'heat haze' was very marked and caused problems for filming throughout the summer. This puzzled us at first, until we realized that the thawed, dark brown tundra absorbed so much more of the sun's heat than did the white, reflective snow. Where two masses of air of markedly different temperatures mix, heat haze develops regardless of the overall temperature.

In summer the tundra became a maze of shallow lakes with boggy areas and drier hummocks in between – the perfect nesting habitat for waterfowl and numerous other species of birds. But for us it meant hip waders whenever we left the immediate camp area. Our standard clothing at this time consisted of long thermal underwear, two pairs of woollen socks, heavy jeans, thick shirts, woollen sweater, down jacket with hood, and the inevitable hip boots that we came to detest after months of constant use. During the first month, gloves or mittens were essential, and we carried lightweight down trousers in our packs. While walking it was fairly easy to keep warm, but filming in one spot for long periods in the icy wind, or sitting still in a blind, we often became chilled. Because of the changeable weather we always carried lightweight rain gear, and even on warm days we never left camp without our down jackets, for a sudden shift in the wind direction in these latitudes can send the mercury plummeting as much as thirty degrees in half an hour. Hudson Bay remained frozen until late July – a gigantic ice block only a few miles from camp!

On the way to our blinds one clear and sunny morning in early June, we laughed to see snow and blue geese slipping on the slick ice of shallow ponds that had refrozen on the surface overnight. Skating first on one foot, then on the other, the geese rarely lost their balance completely although they occasionally broke through the thin ice. Skating didn't come as naturally to a nearby male willow ptarmigan. His solution whenever he slipped and lost his balance was to let his rump touch down on the ice for extra support – an extremely difficult manoeuvre while trying to retain his composure!

3 Filming Begins

Les had the task of 'seeing us into' the blinds each day and returning for us up to twelve hours later at prearranged times. We would approach slowly to give the goose ample time to cover her eggs before she walked a short distance away to join the gander, then one or two of us would enter the little tent and quickly set up tripod and cameras before the remainder of the group walked away. Satisfied that the danger had passed, the goose would soon be back on her nest and we were able to record the natural behaviour of the birds from our concealed position.

Waiting quietly in blinds for hours on end requires patience and can be most uncomfortable. Using a tiny folding stool to sit on, we certainly needed our warmest clothing, especially if a cold wind happened to be blowing though the small opening where our camera lenses protruded from the front of the tent. In boggy areas, water often seeped through the floor of a blind, but if the floor was dry a down sleeping bag pulled up over our legs and lower body helped to keep us warm, while leaving us free to operate the camera. To help ward off the cold we usually took with us a little food and a Thermos of hot water for tea or coffee. It is impossible to heat a photographic blind, even if one has the facilities, for the warm air escapes through the opening for the lenses and causes a shimmering heat haze as it meets the cold outside air, thus ruining any chance of photography. However, despite the chilly conditions, we love working from a blind as it gives us a privileged, close-up look into the private lives of the birds, a view which would otherwise be utterly impossible. There is no time to be bored, as we have to remain constantly on the alert to film any interesting activity, and the unexpected can happen at any moment.

When Les came back for us, we were usually warned of his distant approach by the reactions of geese nearby, and unless we were filming something particularly fascinating we signalled him to come up to the blind. This signal, which he could spot through binoculars from a distance, consisted of hanging a small coloured cloth out of the back door flap of the tent. In this way our presence caused the least amount of disturbance to the nesting birds. Jen and I normally worked from different blinds, comparing notes at the end of each day about what we had seen.

After leaving us at our respective blinds, Les usually spent the day moving other blinds – we gave him the title of 'Chief Blind Putter-Upper' – and searching for the nests of the various birds that are summer residents on the tundra. As the scientists also let us know when they found nests of any of the rarer birds, it wasn't long before our little orange marker flags showed the location of an amazing variety of active nests on the seemingly bare and inhospitable tundra. Even more surprising was the fantastic camouflage of most of the nests, for these ground-nesting birds made clever use of the small amounts of vegetation available on the open terrain. Because of the very severe winter weather conditions, all the bushes are tiny and ground-hugging. Willows and birches on the tundra are usually only six inches high, with the branches spreading out horizontally rather than growing upwards. The tiny bushes mingle with lichens and mosses, as well as short grasses and sedges, to form a springy covering to the constantly damp ground. This is 'dry' tundra, pockmarked with shallow pools of every imaginable size and shape; they in turn are dotted with innumerable islands, ranging in size from less than a square foot to large sections of tundra supporting more ponds. Although the whole area is flat, small connecting channels link up the pools and lakes, and seepage moves the water towards the river and out into Hudson Bay. Many of the birds nest on islands for greater protection against weasels, Arctic foxes and wolves, although these mammals can all swim well. Sadly, the wolves are becoming increasingly rare in the north, mainly because of the Eskimo's new mobility in winter in snowmobiles. We saw only one during our entire summer at the McConnell, and it was running away as fast as it could.

As well as the blind at our original snow goose nest, we soon had blinds close to a pair of blue geese, and a mixed pair where the male was a snow and the female a blue goose. Before the summer was out we had photographed nesting Canada geese, Ross's geese, king and common eider, oldsquaw, pintail, greater scaup, Arctic and red-throated loon, Arctic tern, herring gull, parasitic jaeger, savannah sparrow, Lapland longspur, snow bunting, northern and red phalarope, golden and semi-palmated plover, dunlin, and two other species of sandpiper – the semipalmated and pectoral.

From our blinds we were able to have our first close look at the snow and blue geese. The plumage of the snows is completely white except for the black primary feathers at the tip of the wings. Eyes are brown and the bill pinkish-red, while the edges of the mandibles are black with conspicuous serrations. This black area on each side of the bill is often referred to as a 'grinning patch'. The legs and feet are also a deep pink, with contrasting black nails. The blue goose differs in having a dark body and wings, the head and neck only being white, though there are varying amounts of white on the belly. However, blue is something of a misnomer as the dark portions of the plumage are really a deep greyish-brown. We noticed that many of the snows and blues had an orange stain on their normally white heads, evidently caused by feeding in iron-stained waters during the winter months in Louisiana.

We were particularly pleased to be able to watch and film the mixed pair of snow and blue geese from the close proximity of the blind. Mixed pairs are not uncommon, although the tendency is for the geese to mate with their own plumage type. In cases where the parents are a mixed pair, each offspring will be either a snow or a blue, with the blue strain often predominating. Scientists have now learnt that the colour is controlled by one dominant gene.

Our mixed pair of geese had their nest a few yards from a small pond, but the white male rarely bathed until the blue female had carefully covered her eggs and left the nest to join him. Walking together to the water's edge, they had a leisurely drink before wading a short distance into the shallow pond. A goose bathing is a marvellous sight. With legs bent the bird repeatedly bobs its head and neck beneath the surface causing rivulets of water to flow along its back each time the head is lifted. Standing upright it gives a shake of the head and ruffles its feathers before lowering itself on to the water again. Crouching low on the surface and beating the water with partly opened wings, it flings a fine spray in all directions, often exciting its mate to splash more vigorously in unison. Then the goose stands erect once more, head and neck stretching skywards, powerful wings flapping briefly to shake off excess water before it wades to shore to begin the elaborate grooming process. By running its serrated bill along one feather at a time, the preening goose keeps the interlocking feather barbs in good order. Watching closely, we were surprised to see that each feather is grasped broad side on so that the long feather barbs are flattened against the shaft as the bill works rapidly from the base to the tip of each feather. By rubbing its head first on the preen gland, situated just above the base of its tail, and then over its feathers, the goose waterproofs its plumage by spreading the oily secretion from the gland. While Jen was photographing this mixed pair of geese on 9 June, she saw them mating in the shallows near the edge of the pond, before carrying on with their usual bathing and preening routine. This was a rare occurrence on the tundra, for during the many hours we spent watching the nesting geese, this was our only observation of mating. This pair already had their full clutch of eggs and the female had been incubating for several days.

One day Lee and I were in a blind near the coast in one of the densest parts of the snow goose colony. In front of the blind I could see two occupied goose nests not far from each other within fifteen yards of where I sat concealed. Above the constant gabbling of the geese we heard a sandhill crane calling on the wing, and I was able to film it flying past quite close to the blind. It landed out of camera range near a goose nest, but was soon chased away by the gander and flew back towards us. Suddenly, from just behind the blind, there was a great commotion from the geese accompanied by loud calls from the crane. Peering through a small peephole in the side of the little tent, Lee whispered excitedly,

'Des, the crane is raiding a snow goose nest!'

Using my pocket knife I hastily cut a slit near the back of the blind, and soon had the movie camera trained on the scene, while Lee held the canvas away from the lens. In the cramped quarters of the blind, this movement of the camera and tripod to film in the opposite direction was not as simple as it sounds. Once I began filming the action it was my turn to give Lee a running commentary.

'The two geese are just standing twenty feet from the nest watching the crane eat their egg,' I reported. 'No, wait . . . There goes the gander. He's chasing the crane away from the nest.'

The goose walked over to the nest and inspected the damage, then proceeded to do something that amazed us. She ate the liquid contents of her own broken egg. Perhaps the goose needed the nourishment, but her main purpose was no doubt to keep her nest and remaining eggs clean, and to remove signs of a freshly broken egg which might attract other predators. Once the eggshell was empty, the goose picked it up in her bill and carried it several feet from the nest before dropping it to the ground. Returning to her nest, she carefully rolled the three remaining eggs with her bill before settling down to incubate. This particular pair of geese became rather special to us, and we are happy to report that they successfully hatched their three goslings. Not every gander would be aggressive and fearless enough to chase off a marauding sandhill crane!

All too frequently we were able to observe the raiding tactics of the birds that were the main predators in the snow goose colony – the parasitic jaegers. Because of the jaegers we tried to spend as little time as possible moving about in the colony, for this inevitably caused some of the geese to leave their nests untended for short periods. Flying back and forth above the nesting geese the jaegers' sharp eyes watch for a nest without a female sitting on it. Alighting on the side of such a nest, the jaeger lifts its head high before bringing the tip of its beak down hard on to a goose egg. After a few powerful blows to break the strong shell, the jaeger begins its feast, but is rarely left alone to enjoy it in peace. Other jaegers are ever on the lookout and before long there are several squabbling over the egg. Many times we saw a harassed jaeger take off with either an egg or an embryo in its beak, only to be chased by the other birds. Higher and higher the jaegers climb until a pursuer overtakes the first bird. A mid-air fight ensues, and the prize is finally dropped from a great height. But long before it reaches the ground it is expertly snatched by another acrobatic flier. Then a new chase begins and the piracy continues, until one jaeger at last manages to swallow the food. Should a herring gull appear on the scene while jaegers are raiding a nest, the gull is immediately recognized as the boss and takes over. Some pairs of geese, possibly the younger birds, seem uncertain of what to do when predatory birds raid their nest. Flying half-heartedly at the invaders they chase some off in one direction, while others double back to continue their feast at the now untended nest. The more experienced and aggressive geese chase the raiders off the nest but then remain on guard close by. Each time that we watched a goose return to find a broken egg at her nest, she behaved in the same manner as the goose whose egg had been opened by the crane: the liquid remaining in the egg was eaten before the empty shell was either pushed or carried away from the nest.

Arctic foxes often prowled through the goose colony, but proved extremely shy of humans and we were unable to film any. The foxes too have a liking for goose eggs and frequently carry them off intact to bury for later use. We found these buried eggs in a variety of locations, even on top of Camp Esker; some of them looked as if they had been buried a very long time.

A few years ago at the McConnell River, biologists had watched a single fox systematically clean out the eggs from two hundred goose nests in a period of five days. In spite of the various predators, it is estimated that only ten per cent of the total number of eggs in the colony are destroyed by them or by foul weather in a normal season.

From 10 to 19 June we had ten days of rain, fog, cloud and wind, with very little sunshine. It was a depressing time, for although there was not much more for us to film in the goose colony until the young hatched, we were keen to work with the many other species of nesting birds. Inevitably much of our filming had to be done when the weather conditions were far from ideal. Apart from the lack of sunshine, the wind proved a major problem, causing the blinds to flap and vibrating our cameras. Because of the almost constant wind, sound recording was often impossible. Winds of between ten and twenty miles an hour were normal, with thirty miles an hour not unusual. In mid-June we experienced the first of three particularly violent storms that hit us during the summer. We were out on the tundra two miles from base when a light drizzle began in the early afternoon, and long before we reached camp we were fighting a thirty-five mile an hour headwind. By evening the wind was gusting above forty miles an hour, blowing the now heavy rain horizontally against our tents. In most places one can hear a gust of wind approaching, but on the open tundra there are no obstructions in its path. The only time we heard the noise of the wind was when gusts actually reached our tent. Then, and only then, would we hear the familiar whistling of the wind in the guy ropes, and the sudden whip-like crack as a gust bowed in the canvas side of the tent momentarily, then sucked it back out again as the pressures changed. Sleep was almost impossible and we wondered how the tent could take such punishment, as the storm continued unabated for thirty-six hours.

During the storm we were unable to visit the snow goose colony, but from past observation we were sure that each female would be sitting low on the nest and facing into the icy wind and rain. Only the bird's bill is not protected by layers of down and feathers, but she remedies this by lying her head along her back with the bill tucked under the feathers of one wing. None of the nesting geese we had been observing appeared to be in any way affected by the storm. Fortunately this severe weather occurred a few days ahead of hatching, while the goslings were still safely enclosed within the protective eggshells, and kept evenly warm beneath the tightly sitting mother. If such a storm strikes the goose colony during the hatching period, many of the new goslings are lost. The severe climate on the tundra nesting grounds is the greatest single threat to the snow goose breeding success in any one year. If the spring thaw is too long delayed by unusually cold weather, many geese will fail to nest at all, or at best, if the delay is not too great, will lay smaller clutches of eggs than normal. They know instinctively that in the short Arctic summer, to begin nesting after a certain date is futile. The reader may be amused at the thought of a bird knowing the date! But the annual migrations of geese and many other birds are surprisingly punctual each year. Even in a normal season, should the first clutch of goose eggs be lost for any reason, the short northern summer allows no time for re-nesting – something that is usually possible for birds breeding in milder climates.

As we mopped up in the main tent after the storm, we suddenly heard strange squeaking sounds coming from a pile of boxes. Searching there, we found that a short-tailed weasel had taken up residence among the boxes of supplies in the outer portion of the tent. Giving a high-pitched 'cheep' it darted from its hiding place and leaped lightly on to a table. To our surprise the weasel was unafraid and provided we moved slowly it allowed us to come within a few feet of where it crouched. Already it had lost the winter coat of white which gives the weasel its other common name – ermine. Now it was brown above, with the underparts a light sulphurous yellow. The weasel had a body about six inches long and a four-inch tail ending in the distinctive bushy black tip. In winter this black tip is the only part of the ermine easily visible as it moves across the white snow. This weasel remained around camp for several days, often running over the snow and ice along the riverbank, where its brown coat stood out conspicuously. As the weeks passed we were to see many more weasels, but all too soon we learnt that pretty though they may be, their actions did not always endear them to us. They are expert swimmers, and one of the weasels was seen swimming across to an island in the river close to camp, where apparently it feasted on the eggs in two of the nests that we were filming there – those of a snow bunting and a semipalmated plover. Unfortunately these were two species whose young we had particularly hoped to film and we had no other chance to do so. In early August a weasel somehow found its way into a well-protected cage housing young Lapland longspurs which the biologists planned to use later for breeding experiments. The bloodthirsty weasel killed many of the sparrow-sized birds, although it could eat only a small fraction of the dead ones.

The male Lapland longspurs were among the first birds to reach the McConnell River in May, the females arriving a week or two later. The territorial songs of the male longspurs seemed to accompany us wherever we went at the McConnell, and we will long remember their musical aerial displays. Snow buntings had been seen each year at the McConnell, but nobody had ever managed to find a nest, as they mainly nest farther north. Usually these neat grass-lined cups are built out of sight among heaps of rocks. We even made piles of stones to try to encourage them to nest near camp, but although one male did carry a few pieces of grass to a rock pile, no nest was ever built there. Later, however, we often saw one pair of snow buntings along the river bank less than two hundred yards upstream from camp. By watching them closely through binoculars, we finally located their cleverly hidden nest in the bank of an island twenty yards out in the river. The female had just begun incubating her clutch of six eggs when the nest was destroyed. Although we had no definite proof that the weasel was to blame, we felt the nest had been so well hidden under the overhang of the bank that no predatory bird even flying close by could possibly have noticed it.

On this same island we had seen a pair of semipalmated plovers and felt sure these attractive little birds had a nest nearby. As with all plovers, their nest is merely a depression in the ground and the four speckled eggs blend in with the surroundings as a wonderful example of camouflage. Each time we approached their nesting territory they turned on a marvellous broken wing display, grovelling on the ground and beating their half-opened wings as they attempted to lead us away from the area. Once they were used to us, the plovers became incredibly tame, allowing us to come within a few feet of the nest without budging off the eggs. After their first clutch was destroyed this pair re-nested a short distance downstream and we kept a close watch on their eggs. One morning in early July, Les reported that the eggs were pipping, but a few hours later we found them smashed and empty, probably the victims of a marauding jaeger. We felt very sorry for this pair of semi-palmated plovers, as they had tried so hard and seemed like old friends to us.

In mid-June Charlie warned us that if we wanted to take pictures of the spectacularly coloured male king eiders, we would have to get a move on.

'I've seen quite a few around Caribou Antler Lakes,' he said, 'and that area would probably be the best place to try.'

These large sea ducks come ashore only during the nesting season and even then the males do not stay for long. As soon as the female begins incubating he returns to the sea, joining other males to form large rafts off the west coast of Greenland. Diving to depths of up to one hundred and fifty feet, they feed mainly on shellfish and crustaceans.

Caribou Antler Lakes consist of a maze of lakes spread out over a large area a mile or more south of camp. We visited the region a number of times and saw several king and common eiders in the distance, but they always flew off before we could come within photographic range. Even placing a blind on the shore of one likely lake brought only mediocre results, and with so much else to film we reluctantly decided we couldn't afford to spend any more time chasing eiders.

There were scattered snow goose nests near Caribou Antler Lakes and Charlie thought he detected the call of a pair of Ross's geese as he walked through the area one day. Smaller than the snow, but similar in appearance, the Ross's goose is a separate species, *Chen rossii*. Nesting among the snows, and outnumbered by a thousand to one at the McConnell River, it is not an easy matter to find a Ross's goose nest!

Returning with Charlie to Caribou Antler Lakes on 21 June we had no luck locating the Ross's goose nest, but we did come across two male king eiders on a small lake. Approaching with extreme caution because of our disappointments in the past, we were agreeably surprised to find that the eiders did not retreat. But we soon realized why: they were both intent on remaining close to a tiny island where a female king eider had flattened herself on the nest, her finely patterned brown plumage blending perfectly with her surroundings to make her all but invisible. In contrast with the dull plumage of the female, the males were indeed spectacular, as their scientific name implies – *Somateria spectabilis*. Although the wings and body were immaculately marked with black and white, it was the head above the white chest and neck that caught one's eye. The crown and back of the head are a pale blue-grey and the white cheeks washed with pale green. The bright orange-red bill is tipped with a pink nail, but it is the large fleshy protuberance stretching upwards from the base of the bill that is so unusual – bright orange and bordered with narrow black lines, it tops off the male king eider's fantastic appearance.

We stayed close to the king eiders for several hours, and by the end of that time were within ten yards. While the males remained on the water most of the time, we noticed that whenever on land they walked almost upright, not unlike penguins. Unfortunately there was a strong wind blowing that day which made filming difficult, even with the others trying to shelter the movie camera by opening their jackets and holding the sides out like bats' wings to act as a windbreak. But we were quite excited at finding the eiders, for with the female already sitting, the males would not linger much longer. A few days earlier we had located another sitting female king eider, but never saw a male anywhere near her nest. The mass of grey down with which the female eider lines her nest is incredibly fine and soft, and is of course the reason why down-filled quilts are called eiderdown.

4 The Week of the Goslings

June 23 proved to be a very exciting day for all of us, as it was then that we saw our first snow and blue goose goslings. So closely co-ordinated is their nesting that within a week virtually all the goslings had hatched, leaving just a few scattered pairs of geese to try to bring off their broods during the first few days of July. These late nesters have a difficult time with the jaegers, for once the majority of the goslings have hatched, the predators concentrate their raids on the few remaining nests.

The first sign of an egg hatching is a 'starring' of the shell towards the larger end. This is caused by the gosling tapping persistently with its pointed egg tooth against the inside of the shell. Located at the tip of the upper mandible, this egg tooth is the gosling's key to life in the outside world. A few more pushes, and small pieces of shell break away, leaving a quarter-inch hole through which the gosling now breathes and chirps. Even while still imprisoned within the egg the gosling has been making clearly audible cheeping and tapping sounds, but now the calls are much louder. With frequent rests, the gosling rotates within the pipped egg, using the egg tooth to make a series of cracks and small holes in the shell. Finally, about twenty-four hours after the first hole appears, the youngster gives a series of energetic heaves, pushing the entire end of the eggshell outwards. Struggling clear of the shell the gosling flops helplessly in the bottom of the nest. Exhausted after its battle to escape from the egg, this wet and bedraggled creature appears more dead than alive. In a few short hours, however, it is dry and fluffy and ready to face the world. It is difficult to imagine anything cuter than this two-and-a-half ounce ball of dark greyish-yellow down, with a halo of bright yellow, and tips of fine, longer down glowing like spun gold in the sunshine. The blue goose goslings are a much darker grey all over, but often sport a small patch of yellow at the base of the lower mandible.

During the hatching period we found that we could approach much closer to the geese before they retreated from the nest. Indeed, a few of the females had become so fearless that they even flapped aggressively towards us. Usually all the eggs in a nest hatch within a few hours of each other and the goslings are ready to leave the nest about twenty-four hours later. Although their yolk sacs will continue to sustain them for the first few days, the goslings begin pecking at anything in sight while still in the nest. Checking to see what is edible, they pull at twigs and leaves, and within a few hours are expertly catching mosquitoes off each other. There is little food for them in the nesting colony, as the adult geese have been living there for weeks, and the parents lead their family away as soon as possible towards the grassy coastal flats. This is known as the brood migration, and the goslings may travel as much as forty miles during the first week of their lives.

As they walk across the tundra the adults move with a high-stepping gait, taking care not to tread on the young, which are often close to their large webbed feet. The parents try to keep their brood close together for safety, but the goslings often pause to peck at greenery or mosquitoes. Racing full pelt to catch up, a gosling sometimes trips, tumbling over and over in its haste, but the laggard is so well cushioned by its thick layer of soft down that it suffers no harm and soon rejoins the rest of the family. Many of the geese use the McConnell River as a floating highway to speed them on their way to the coast, saving many miles of walking. Communicating all the time with their brood by uttering encouraging little grunting sounds, the adults lead the goslings down the river bank. Seeing their parents wade into the water the youngsters hesitate only momentarily before following, although they may often have to jump several inches from the bank into the water. The adults head out into midstream and remain side on to the current with the young ones sandwiched protectively between them as the river takes the little flotilla down small rapids, with the buoyant goslings bobbing like corks. We watched one family become temporarily split in the fast flowing rapids: the goslings all grouped together on the upper side of a large rock in midstream, while both adults were swept a short distance downstream. Four of the babies swam around the side of the boulder in the strong current, while the fifth one decided to climb up over the top of the large rock rather than swim. He then jumped down into the water again, bobbing off downstream to rejoin the rest of the family. Occasionally a straggler isn't so lucky and falls prey to a marauding herring gull which swoops down with open beak to snatch the gosling from the river. When approaching larger rapids, the geese were careful to shepherd their young family towards shore, walking safely along the river bank until well below the danger.

The peak of the snow goose hatch was on 26 June. It was the first really pleasant sunny day we had experienced on the tundra; warm and without a strong wind. After spending most of the morning filming families of geese crossing the tundra and floating downstream, we stretched out in the sun on a dry patch of tundra above the river. What luxury it seemed to us to be without down jackets and waders while we ate our picnic lunch! Other creatures were also reacting to the warmth of this sunny day: the mosquitoes emerged in force.

Later that day we checked the nest of a sandhill crane that we had been photographing from a blind for the past two weeks. One of the two eggs was pipped! Jen stayed in the blind to watch the cranes, while Lee, Les and I moved on to a dense part of the goose colony near the coast. We spent a few hours filming the new goose families, then headed back to pick Jen up on our way to camp. As we approached the crane blind around 8 p.m., we noticed a fight going on between the two cranes and a pair of snow geese. The combatants took no notice of us as we hurriedly took off our packs on a dry hummock.

'Can you set up the tripod quickly, Les?' I asked urgently. Meanwhile, Lee helped me unroll the sleeping bag from around the 16mm Arriflex camera.

'The action is going to be over before we're ready!' Les predicted, as he hurriedly tightened the tripod legs.

Fortunately, it did not turn out that way. I was able to film the dramatic action in detail, while Jen was busy taking still photographs from inside the blind sixty feet from the combatants. Later she was able to tell us how the encounter had begun. One moment all was peaceful on the tundra, with the female crane sitting tightly on the nest, its mate nowhere in sight. Then without warning the sitting crane suddenly left the nest, calling loudly as it ran towards a family of snow geese twenty yards away. With their four tiny goslings, the geese were merely passing through on their way to tiny river and posed no threat to the cranes. However, as one of her own chicks was now hatching, the crane was evidently in a very protective mood and would tolerate no trespassers. Alerted by the loud alarm calls of its mate, the second crane flew in to her aid. It was at this point that we came on the scene and I began filming.

The cranes ran at the intruders repeatedly, causing the geese to bunch together for safety. Flying a few feet in the air, the larger male crane struck savagely at the geese with long legs and powerful feet. The courageous gander held his ground between his family and the cranes, often flying up to meet the crane in mid-air in an effort to protect his brood. One colour picture Jen took shows the crane's foot striking the gander's head forcefully during one mid-air clash, the impact of the blow bending the head and neck of the snow goose back along its body. Apparently this caused no harm, as the gander continued to defend his family just as vigorously as before. While filming the action I saw a crane actually pick up a tiny gosling in its bill and toss it high in the air! The gosling was so light and downy that it merely bounced on the springy tundra, before running to rejoin its family. Undoubtedly a very frightened little gosling, but otherwise unharmed. Gradually the geese moved their young farther away from the cranes' nest and the fight was finally over. We had been so busy filming that we had no idea how long the conflict lasted.

'Did you get all that action?' I called to Jen as we neared the blind.

'I don't know,' she replied. 'But every camera I have in here is out of film!'

We left the area quickly, as the cranes were strutting back and forth calling loudly, obviously still agitated after their fight with the geese. It had been a very long and tiring day, full of excitement for us with so many goslings hatching out in such beautiful sunshine. The crane-goose fight was totally unexpected and something no one else had ever seen, as far as we knew. Our aching muscles almost forgotten, we trudged back to camp just as the sun was dipping like a fiery ball towards the horizon.

Although the next day was cold and foggy we returned to check on the cranes but found no change in the eggs; only one was pipped. As there was almost no wind – a rare event at the McConnell – we made sound recordings of the cranes calling in protest at our presence and then played this back to check on the quality of the recording. At this the pair of cranes became very agitated, flying and running around near us as they searched for the invisible cranes that had dared to invade their territory! Once again we beat a hasty retreat to allow the cranes to settle down to their domestic chores in peace.

Our notes for the following day, 28 June, record that one chick had hatched and dried, while the second egg was now pipped. Both chicks were making high-pitched piping sounds, one calling from within the egg. This was late June, just past the longest day of the year, and yet it was bitterly cold with drizzling rain, and a below freezing temperature of only 30°F at 11 a.m. We remained near the crane nest for a few minutes only, before moving more than a quarter of a mile away to film Allan Aubin, one of the graduate students, web-tagging goslings still in the nest. These tiny metal tags are placed on the webbing between the gosling's toes and are replaced with permanent numbered leg bands if the same birds are caught a month later during banding drives.

When we returned to the crane blind to store some equipment there before heading back to camp, we were concerned to see that both cranes were some distance away from the nest. On checking, we found both the chick and the egg were very cold, and cursed ourselves for disturbing the cranes too much. While Jen hurriedly packed away our equipment I put the chick inside my shirt where it would absorb the warmth from my body, and Les did the same with the egg. When they were both fully warmed we returned them to the nest and hurriedly left, relieved to see the pair of cranes standing nearby.

As we returned to check on the cranes next day we were all feeling rather subdued, and anxious to know what awaited us in the nest. The reddish chick was wandering about, one parent was some distance from the nest, and loud pipings still came from the second egg. The second adult crane left the nest briefly when we approached, but did not stay away for long. From the concealed position in the blind we could watch both the incubating crane on the nest and its mate wandering across the tundra searching for food for its chick. We were surprised to see the foraging crane raid a blue goose nest only fifty yards from its own. Carrying a still wet, dead gosling in its bill, the crane dropped this close to its own chick. After feeding pieces of the gosling's yolk sac to its chick, the crane devoured the remainder of the victim itself. Although we had spent many hours in this blind, this was the only time we saw one of the cranes raid a goose nest anywhere near their own nesting territory. During the incubation period, a truce seemed to exist with their neighbours. One crane would sit quietly and inconspicuously on the nest, while its mate flew half a mile away to feed, still within calling distance of the crane at the nest. However, once the cranes had hungry young of their own to feed, it seemed that anything and everything was fair game.

Reading our notes for 30 June, we recall that 'today we found both baby cranes about three hundred yards from the nest, each parent taking care of a separate chick. The adult nearest to us showed its displeasure at our presence by picking up pieces of turf and mud in its bill and throwing them high in the air, but it soon calmed down and took no further notice of us as it carried on looking for food. Suddenly a semipalmated sandpiper flew up from almost underneath the crane's feet, giving away the location of its nest. One by one, the crane daintily picked up the four tiny eggs in its bill, swallowing each one whole. A short time later the second crane located a dunlin's nest, and taking an egg in its long straight bill, carried it off to feed to the chick.' The next day the cranes moved some distance away and were difficult to locate, so we were unable to keep up our daily observations.

Though the hatching of all the goslings within a single goose nest is usually closely synchronized, it does sometimes happen that the last gosling to hatch is still too weak to follow when the slightly stronger ones begin to wander from the nest. When the parents leave with the rest of the brood, any defenceless gosling left behind in the nest is almost sure to fall prey to a herring gull.

Before travelling north I had obtained a permit from the Canadian Wildlife Service to collect some of the goslings which don't stand a chance of survival if left in the nests on their own. On 23 June we picked up the first stray. Watching a pair of snow geese herding their four youngsters towards the river, we suddenly noticed a fifth one tottering along well to the rear, unable to keep up. While we watched, herring gulls began circling overhead, prompting Les to run in to rescue the straggler before the gulls could swoop down. Placing the exhausted gosling inside his shirt, Les soon reported that the little fellow had quietened down and appeared to be sleeping contentedly. By the end of the hatching season, we found ourselves with a family of fifteen goslings – eleven snows, three blues and a Canada goose. Having known their own parents only briefly, the tiny goslings quickly accepted us as their 'real' parents – a process scientists call imprinting. In fact, we often wondered just who adopted whom!

It is quite easy to imprint many newly-hatched birds on artificial parents, especially goslings. Normally, when hatched and raised by their true parents, it is during the first day or two of their lives that the young become imprinted on the features by which they will be able to recognize and follow the adults. In some species of birds the early process of recognition is far more simple than it is for other species, and obviously the simpler the process the easier it is for a bird to become imprinted on an object other than its parent.

Everywhere we went the goslings tried to follow closely at our heels, just as they would have done with their real parents. For the first week or two they were too slow to keep up as we hiked for miles across the tundra, so we carried them in a cardboard box. Whenever we paused to film, the goslings were given their freedom, but stayed close to us as they ran busily back and forth feeding on greenery, miniature wild flowers and insects. They quickly became experts at picking mosquitoes off our clothing and could even catch one in mid-air. We did not appreciate the hordes of mosquitoes that appeared with the first hint of warmer weather, but these insects form an important food supply for the many species of young birds on the tundra. After half an hour of frenzied feeding activity, the goslings were temporarily exhausted and flopped down in a heap, huddling together for warmth as they rested close to us. While our imprinted goslings were little they were always around our feet and we had to be extremely careful not to step on them inadvertently with our heavy boots. However, they grew so rapidly that this danger soon passed.

Even when they were small, the goslings loved to dive and to swim underwater, but at first they were so buoyant they had difficulty in submerging completely. After a long time in the water they resembled drowned rats, with their once fluffy down plastered to their bodies. We noticed that they seemed to become wetter than wild goslings would have done after a similar swim. When tiny, the wild goslings are regularly brooded under the warm feathers of the mother and presumably their down then becomes sufficiently waterproofed from the repeated contact with the well-oiled plumage of the mother goose. This was one service we couldn't render to our adopted family, and during the first week or two we had to be careful they didn't become wet on very cold days. But it was not long before they were able to preen and use the oil gland above the base of their own tails, and then we had no further worry about their getting too wet.

As we walked along followed by the goslings, we tried to imitate their high-pitched sounds. 'Peep. Peep-peep,' we called. As the days passed, our calls gradually changed until we found ourselves saying

'Creep. Creep-creep. Come on little Creeps!'

And so it was that our goose family became known as 'The Creeps'.

During the first week or two the Creeps needed special care to ensure that they were all feeding well and that they didn't become chilled. On cold, damp days, Lee remained near camp as 'Mother Goose' while we were away filming. At night the little goslings always slept in a box in the warm cabin. Of course the wild goslings are able to snuggle under their mother for warmth whenever they are cold or tired, but it was a little more complicated for us to regulate the temperature for our Creeps. Although they were eating natural foods on the tundra each day, we also fed them on Miracle, a Canadian brand of high protein dog food crumbles, which helped to satisfy their voracious appetites. The first few goslings we adopted had to be encouraged to eat this dry food, but within a day or two they were gobbling it greedily. As new goslings were added to the family they merely followed the example of the others where food was concerned, and it was surprising how frequently their food dish needed replenishing each day.

On 28 June we had an unexpected addition to our growing family. It was cold and windy, with drizzling rain falling and the temperature hovering near the freezing mark all day. After spending most of the day out on the tundra checking on the various nesting birds, we returned to camp earlier than usual to work on notes and correspondence. A short while later, Charlie MacInnes strode into the cabin, wet and chilled after a long hike in search of Canada goose nests south of camp.

'Hey, I have something for you,' he said.

He had a mysterious look on his face, but did not keep us in suspense for long. Smiling, he reached into his jacket pocket and lifted out a reddish-orange ball of fluff with large eyes, a rubbery pointed beak and long thick legs. A tiny sandhill crane chick! He was such a cute little fellow that he immediately became a favourite with everybody, and was to be a spoilt bird forever more. Having had experience with raising crowned cranes in Africa, our thoughts turned right away to the urgent problem of finding the variety of food sufficient for the growing chick.

'I don't know if you're doing us a favour, Charlie,' I said with concern in my voice. 'How on earth are we going to catch enough insects to feed it when we're all so busy?'

'I couldn't leave it on the tundra,' Charlie replied. 'It was wet and stiff with cold when I picked it up. The parents were a quarter of a mile away both tending another chick.'

Half-frozen and almost dead when Charlie picked it up, the crane chick had quickly revived in the warmth of his pocket on the long walk back to camp. And so it was that Red, the baby sandhill crane, came into our lives. Little did we realize he would remain a close member of the family for the next two years. As he grew up, his rusty colouring became greyer and his name was changed to Fred, although to this day we still aren't certain that 'he' is a male!

As we had expected, finding enough insects for Fred was a problem for the first few days and he hungrily downed every one that was offered to him. We all spent hours turning over rocks along the river bank in search of stoneflies and insect larvae. Fred even enjoyed eating the one-inch-long crane flies, so named because of their extremely long legs. Charlie suggested mincing fresh fish caught in the river, and this helped tremendously with Fred's diet in the early days. Gradually, Miracle dog food crumbles were added to the ground-up fish, and this mixture was fed to the baby crane using long tweezers that no doubt resembled its parent's bill. In this way Fred soon developed a taste for the dog food and would eat it dry, straight from the supply that was always available for the greedy goslings. The latter shovelled it into their bills with gusto, beating a track back and forth between the food and the water dish as they washed the dry crumbles down. We quickly learned that it was a mistake to place the two containers side by side. This had been necessary at the very beginning when they were learning to eat this food, but later, if the two dishes were left close together, the food became a useless soggy mess in no time at all and had to be cleaned out repeatedly.

What a contrast it was to watch Fred's dainty eating habits in comparison to the gluttonous goslings! Stabbing his pointed bill into the can of food, he picked up small pieces. Then, by raising his head slightly and giving it an almost imperceptible forward jerk, he moved the food far enough back in the bill to be able to swallow it. Stab, jerk, swallow. Stab, jerk, swallow. These movements continued until Fred's hunger was satisfied, with only an occasional brief pause to look around him. Not until the meal was over did he bother to have a drink.

Growing up with the goslings, Fred at first seemed convinced that he was one of them, and that we were his parents. Later, as he grew taller and began to tower above the young geese, there was no doubt in our minds that Fred was sure he was a human being!

Everywhere the Creeps went, Fred went too. While they ate a variety of greenery on the tundra, he was unable at first to find much food for himself. So we soon learned to take a supply of Miracle with us whenever we expected to be away from camp for more than an hour or so. When the Creeps huddled together for warmth on chilly days, Fred often lay down in their midst, his rusty colour looking strangely out of place amongst them. Sometimes he would keep going until he was exhausted, then collapse and stretch flat out on the tundra beside us. The only time the Creeps were silent was when they slept but Fred seemed to keep up his high-pitched little purring noises even when asleep! We soon learned to mimic this sound, and when uttered within his hearing it always prompted an immediate response.

The Creeps were real water-babies, but we doubt whether baby cranes normally swim for pleasure, even though they spend much of their time following their parents through swampy areas in search of food. Little Fred quite enjoyed wading in the shallows and occasionally even dipped his head right under the water to wash it. The Creeps usually headed for deeper water where they could splash and dive to their hearts' content. Although inadequately equipped with his unwebbed feet Fred soon learnt to follow, in a vain attempt to keep up with the goslings. When just over a week old, Fred surprised us by having his first real bath, immersing his head and neck repeatedly and rolling first on one side, then on the other, in order to get water right over his back. He never matched the Creeps' love of the water, but he usually bathed about once a week, and continued to go swimming with them.

5 Life Explodes on the Tundra

During the short Arctic summer of continuous daylight the tundra explodes with life. Many of the nesting birds have migrated thousands of miles to reach their summer home. Some, like the snow and blue geese, come from the extreme south of the United States along the Gulf of Mexico; others, such as golden plovers and some of the sandpipers, have flown much farther from the pampas and coastal flats of southern South America; but the Arctic terns are the champion migrants, having spent the austral summer in Antarctic regions.

Regardless of the species, the nesting is timed so that there will be an abundance of food available when the chicks hatch. By the end of June and early July, the weather has begun to warm up. Wild flowers bloom in profusion, insects appear, the tundra ponds teem with minute animal life; in the clear waters of the shallow lakes and ponds myriads of daphnia, mosquito larvae and fairy shrimp can be seen. Viewed from above, the delicate fairy shrimps resemble one-inch-long transparent fish. The pond bottom provides a hiding place for other aquatic life, such as the prehistoric-looking tadpole shrimp. Large bees visit the tiny wild flowers and a few small butterflies flutter at the mercy of the wind, sailing across the tundra only a foot or two above the ground.

It was mid-June before we noticed the first wild flowers – the tiny white bell-shaped flowers of the bearberry, which hug the tundra closely in the drier areas; but it was early July before the flower display reached its peak. The most prolific flowering plant was the Andromeda, or bog rosemary, which began to colour the tundra in late June and continued to bloom for a month. These small flowers clustered together in such quantities that they turned patches of the tundra pink. It was now that we noticed the many tiny bushes of Lapland rosebay, a member of the rhododendron family. In Australia we had been familiar with the colourful displays of eight-foot-high azalea and rhododendron bushes in bloom. On the tundra the miniature rhododendron blossoms were the same deep pinkish-mauve but the bushes themselves were a scant six inches high, with branches hugging the ground as they spread out horizontally from the main stem. Blooming in profusion on the drier parts of the tundra were the creamy white mountain avens, the official flower of Canada's North-West Territories, while buttercups and bladderworts pushed their flowers above the surface of shallow ponds. There were even miniature orchids nestling inconspicuously among other plants, visible only to the sharpest eyes. During the short flowering season we photographed almost forty different species of wild flowers. It was a tremendous help to be able to identify each one accurately, thanks to the work done in this area during previous years by Charlie's wife Kaye. By the time the goslings emerged, the tundra was a Lilliputian fairy garden.

Of the countless swarms of small insects it was the mosquitoes which made their presence felt most of all! Except when strong winds blew, they were with us in hordes from the latter part of June until the end of August. Their bites were not too painful for we always carried a supply of insect repellent, and in the end we had been bitten so often that we became partially immune, but having clouds of them flying around one's face and into eyes, nose and mouth was no joke. On bad mosquito days we could not hand hold a microphone when recording bird calls, as the 'mossies' would buzz noisily around the microphone. If we used a parabolic reflector, dozens of them flew against the aluminium surface, producing an annoying series of metallic pings on the recording.

Our sleeping tents were insect-proof and we were always careful to kill all the mosquitoes that entered with us. On calm nights we often thought we could hear raindrops splattering on the canvas, only to realize the sound was made by hundreds of mosquitoes hitting against the outside of the tent. Our photographic blinds proved to be very popular with these voracious insects, and they came in by the hundreds to shelter from the wind. We found that a mosquito's long proboscis could penetrate one layer of clothing, and we therefore often wore long underwear even on warm days. We were not the only ones bothered by these pests. Caribou would move out on to the bare tidal flats in order to escape them, for mosquitoes settle on the tiny birch and willow bushes when it is too windy or too cold for them to fly and then rise in dense clouds whenever an animal walks past. Though birds are protected by their feathers, mosquitoes cluster at the base of their bills and around their eyes, searching for any bare skin through which they can gorge themselves on blood. We felt especially sorry for birds sitting patiently on their nests, for all they could do was to shake their heads repeatedly, or wipe the mossies off against their back feathers; even then, the hungry insects soon settled again. However, when the young birds hatch, mosquitoes provide an important source of protein for the chicks.

Luckily for us there were far fewer blackflies than mosquitoes. These tiny insects have a nasty habit of crawling unnoticed under one's clothing, where they chew out a small piece of flesh without hurting at the time, but the resulting bumps remain sore and itchy for a week or more. We saw far more caribou at the coast when the blackflies became numerous in the interior.

By early July the tundra was strangely silent. From our camp we could no longer hear the constant murmur of activity from the goose colony across the river. On a walk of two or three miles across tundra that less than a week ago had been densely populated by snow and blue geese, we now saw only half a dozen pairs of late nesters. It was as if the many thousands of hatched goslings, together with their parents, had disappeared into thin air. They were now spread out down on the coastal flats, feeding on the plentiful short grasses, and we rarely saw any of them during the next two weeks. On the flat terrain they could see an approaching human over a mile away, and kept well out of our path. Though we couldn't film the geese during this period, we had our hands more than full working with the many other species of hatching birds and looking after our adopted family.

The two things that surprised us most about the majority of tundra-nesting birds were their amazing ability to camouflage a nest on the completely open terrain, and the fact that most species showed no fear of man. Perhaps the shyness shown by geese and other game birds could be partly attributed to the hunting pressures they face at certain times of the year; most species nesting on the tundra are not hunted at all. Only in the Galapagos Islands, where man has hardly interfered with the birds and animals, had we met with such trust before.

Because of our daily visits, the female ptarmigan near our snow goose blind became particularly tame. She would not fly off as we approached slowly, and Jen would have to lift her gently off the nest in order to check whether her eight eggs had begun to hatch, as we did not want to miss the opportunity of filming her chicks when they appeared. Standing fluffed up beside her nest, the female ptarmigan clucked indignantly while we inspected the eggs, returning to warm them again as soon as we began to move away. Late on 3 July her eggs began to hatch, and by sundown four of the chicks were out, but there was not enough light for filming close-up scenes. Next morning we waited impatiently in camp while dense fog prevented us from going across the river to photograph the ptarmigan chicks. Finally, around two o'clock in the afternoon, the fog began to lift. We arrived to find the ptarmigan nest empty, but still littered with empty eggshells. We walked in systematic, ever-widening circles around the nest and before long found the ptarmigan family only fifty yards away. All eight chicks had safely hatched and were with the parents. There can be hardly any baby birds more appealing than these willow ptarmigan less than twenty hours old. Complete with feathered legs and feet, the tiny chicks were attractively patterned with shades of brown above, their undersides being pale yellow. Moving about the tundra, the adults ate bog rosemary and avens flowers as well as greenery, while the chicks followed, pecking hungrily at everything that caught their fancy. Noticing a gull approaching on the wing, the female ptarmigan gave two or three staccato calls. The chicks knew instantly what to do, flattening against the sides of small hummocks or under tiny rhododendron bushes, their markings rendering them invisible while they waited without moving a muscle. When the danger had passed they all started feeding again. On one occasion, when the female gave her alarm call, it took us two or three minutes to locate the cause of her concern – five jaegers flying so high we could barely see them.

When she felt the chicks were becoming tired or chilled, the female ptarmigan crouched on the tundra and called her brood to her with a series of low clucking sounds. The scattered youngsters gathered about her, pushing and shoving to get under her warm feathers. Occasionally a chick climbed on to her back, only to slide slowly to the ground at her slightest movement. As the chicks livened up again after their short rest, tiny heads peeped out from beneath her wings, until the mother tired of their wriggling and stood up, leaving a heap of fluffy chicks behind as she moved off, feeding. The young ptarmigan were soon on their feet and following their mother again. The male mostly stayed a little apart from the group, constantly on guard, and although a chick or two sometimes followed him, he never brooded the young. After this first day, the ptarmigan family moved far from the nest and we were unable to find them again. However, we did have an interesting experience with another family of willow ptarmigan.

Returning to camp one evening, we were mystified by an incessant high-pitched piping coming from just across the creek to the west of camp. Evidently Charlie had been working in camp all afternoon and had heard the plaintive calls several times, but had been too busy to investigate. Wading across the stream, we soon located the source of the strident piping. It was a single willow ptarmigan chick alone in the nest. The others obviously had left the nest before this one was sufficiently dry and strong enough to accompany them, and now it was lonely and cold. We searched the area, but there was no sign of other ptarmigan anywhere around. Once they have left, ptarmigan – like geese – never return to the nest, spending each night in a different place. As it was likely to be a clear and cold night, we reluctantly took the little bird into the cabin for warmth, although we had scant hope that we would be able to raise him. Tiny ptarmigan chicks are precocious from the time they hatch, feeding themselves as they follow their parents across the tundra, and needing only protection and warmth from the adults. Warmth was easy to arrange for the orphan, but a constant supply of food was more difficult. Holding him up to the inside of the cabin windows, where dozens of mosquitoes and blackflies had gathered, he hungrily gobbled up all we put within his reach. He would also need to eat the buds of tender young tundra plants, but that could wait until next morning.

As usual, we were up early and attending to the baby goslings soon after 5 a.m. The orphan ptarmigan was lively so we again searched for the rest of his family across the creek from camp. Les finally found a ptarmigan group with tiny chicks about three hundred yards from the empty nest. Were these the right ptarmigan, and would they accept the little straggler who had been with us overnight? We approached cautiously and set the tiny chick down about thirty feet from the others, then moved away to watch. Finding himself alone once more in the big wide world, he began his plaintive pipings again. The mother ptarmigan immediately raised her head alertly and ran over to herd the orphan into her group. We heaved a big sigh of relief and happily left him in the care of the adult ptarmigan, knowing he would have an excellent chance of survival.

Early in June, a pair of Canada geese began spending a lot of time near camp, apparently quite unafraid of humans. They were leg-banded with numbered metal rings and Charlie felt certain that they were from a group of goslings raised in captivity at the McConnell River two years previously and released at the end of the summer to fly south with the wild geese. From band returns it was already known that many of this group had been shot on their way south during the hunting season, so we were glad to find at least one pair that had survived. This pair was dubbed the 'Miracle Hounds' because of their liking for the dog food on which they had been raised two years earlier, and which was now put out daily for their benefit. We felt sure they would nest some-where nearby, and later in June their nesting site was discovered on a tiny island in a lake only a quarter of a mile from camp. These small northern-nesting Canadas are called Richardson's Canada geese, *Branta canadensis hutchinsonii*. Each time the geese visited camp for Miracle they were as friendly as ever, but if anyone attempted to approach their nesting site, the gander was anything but hospitable. With a human still twenty yards or more away, he flew into the air to attack, making vicious hissing sounds and causing the intruder to duck smartly and throw up a protective arm. We admired the way he protected the nest. Having filmed his aggressive behaviour once we stayed away and left the pair of geese in peace.

Shortly before the Miracle Hounds' goslings were due to hatch, Charlie showed us two neck bands he had prepared for the adult geese. Made out of lightweight aluminium, these were each an inch and a half in diameter and two inches long. They were open at one side to allow them to be fitted around the neck of a goose before being gently closed. Each neck band was clearly marked with coloured tape and a pair of one-inch-high black letters. By using powerful telescopes, biologists are able to make positive identifications of individual marked geese anywhere on the flyway.

Charlie did not have too much difficulty in catching the aggressive gander in a hand net and soon neck-banded it before setting it free again. But the wily female eluded him for several hours and after watching the activity for a while we went off to film elsewhere. At supper that night we asked Charlie if he had managed to catch the goose, and everyone began laughing. It turned out that Charlie had used different neck bands on the Miracle Hounds: one labelled JEN for the goose, and another marked DES for the gander!

Less than a week later the four Canada goslings hatched. Their deep yellow down seemed even brighter than that of the snow goslings as they nestled alongside JEN and pulled at nearby grass. One, more adventurous than the rest, moved to the edge of the nest and tumbled down the bank of the tiny island. Landing on his back with feet thrashing the air, he stopped just short of the water. Although only a few hours old, he had no trouble in finding his feet and returned unaided to the nest. The gander DES stood guard in the shallow water nearby, but his aggression had diminished with the passing weeks and he no longer considered us a threat. Late in the afternoon the female stood up on the nest for several minutes, then moved down into the water, quietly 'talking' to the goslings as she did so. One by one the four goslings scurried over to the side of the island nearest her. Finally one plucked up courage and tumbled down the bank to the water two feet below. The remaining goslings hesitated briefly before following one by one. They were completely at home in the water, swimming and trying to dive within the first two minutes. JEN and DES let their happy offspring enjoy their water games for several minutes before slowly swimming off with them across the pond. It was a beautiful sight to end our day.

While filming the Canada geese, we noticed a large white bird landing on the tundra half a mile away. This was our first sighting of the beautiful snowy owl in the land where it really belongs. We had been hoping to film the snowy owls nesting, but unfortunately for us this was a poor lemming year (lemmings are their main food) so the owls were not nesting anywhere in the McConnell River area. We followed this owl across the tundra for two hours, but it would never allow us to approach closer than seventy yards before flying off for a short distance. Once it alighted not far from nesting Arctic terns, which immediately dive-bombed the owl, causing it to duck its head repeatedly. It soon tired of their aggressive attacks and flew off across the tundra again.

In time, many of the nesting birds, particularly those near camp, became used to our presence and movements. We found that we did not always need a blind, as we could gradually move closer with the tripod while filming and after a few hours work only a few feet away from most nesting sandpipers and phalaropes, without causing them to leave their nests. This is a good illustration of the way most tundra-nesting birds do not associate humans with danger to themselves or their young. Patience is required, but after I had spent a few hours near the nest, one sitting male northern phalarope allowed me to move the Arriflex camera to within three feet. Three chicks had hatched safely and were now lively little speckled balls of yellow, brown and grey fluff running around on long legs, while the fourth egg was pipped and the chick still imprisoned within its shell. As the adult left to carry a piece of eggshell away from the nest in his bill, the three active chicks followed, each busily pecking at tiny insects as they scrambled through the short tundra vegetation. We were witnessing something that must happen frequently each summer: the one late chick is left behind in the nest to perish.

'Des, do you think I should try picking up the babies and bringing them back to the nest?' Lee asked, concern in her voice.

'I doubt if they'll stay,' I replied, 'but it's certainly worth a try.'

Moving slowly, Lee was able to pick up the nearest baby gently and return it to the nest. By the time she returned with the second little phalarope, the first was already several feet away again. The lively chicks just did not want to stay quietly with the fourth still inside the egg.

'I'll try cupping my hands over the nest while you gather the other chicks,' I said.

As soon as the three adventurous chicks were safely back at the nest, Lee took over guard duty while I returned to the camera. Meanwhile the adult male phalarope scurried about on the ground only a few yards from us, then moved closer to Lee's cupped hands as she 'brooded' the three chicks with the pipped egg in the nest. When he was only a few inches away, 'talking' to his family inside the sheltering hands, Lee gently parted her fingers; and as her hands slowly opened, the adult phalarope entered to brood his babies, completely ignoring Lee's close proximity as his ruffling feathers brushed against her fingers. It was fascinating to see a wild bird acting in this fearless manner. Very slowly, so as not to frighten the sitting phalarope, Lee withdrew her hands and backed away.

'Before we leave, you must see what they look like from this angle,' I said.

The male was sitting facing me, and on either side of his body we could see the legs and feet of the chicks as they stretched high, with heads and bodies completely hidden under his protecting wings. It was a charming domestic scene, but with the sexual roles reversed, as it is with all three species of phalarope. The more colourful female phalarope assumes the active role during courtship, then after laying four speckled eggs in the nest, leaves the male entirely on his own to incubate the eggs and look after the young.

Next morning we returned to the nest and were pleased to find it deserted. It was not long before we located the male phalarope busily looking after his family of four chicks as they ran about feeding in the small pools close to the river bank. There was absolutely no way we could tell the chicks apart, as they were all equally strong and active: the 'runt' of the family had caught up.

We located many nests of the small semipalmated sandpipers, always well camouflaged near tiny bushes on the open tundra. By keeping a close daily check on these nests, we were able to spend several hours at one nest when the four eggs were hatching. When we arrived at 6 a.m., one semipalmated sandpiper chick had just emerged. Wet and bedraggled, it lay draped over the other three speckled eggs in the small cup-shaped nest depression in the ground. By mid-morning two more chicks had hatched and we had a clear view of the nest interior each time the sitting bird removed a large piece of eggshell. Gripping the empty shell firmly in its bill, the sandpiper flew twenty yards or so from the nest, dropped its load, and immediately returned to its clutch. The mate, which was slightly smaller and paler in colour than the sitting sandpiper, was not much in evidence until the third chick had hatched. Then it repeatedly approached the nest like any proud parent anxious to have a first look at its offspring, but the sitting bird jealously guarded the brood and always chased the other sandpiper away, even at times flying after it.

Soon, the first-hatched sandpiper chick began wandering short distances, and when it was some six feet from the nest the smaller parent finally managed to take over the care of this isolated chick and sat brooding it on the tundra. Furious at this turn of events, the larger sandpiper left the nest and tried to claim the wandering youngster again, but this time its mate was not to be intimidated and refused to budge. Then a strange thing happened. After so strenuously rejecting all efforts by its mate to come anywhere near the nest, the larger parent suddenly lost interest in the care of its young. It moved off, never again coming closer than twenty feet to the nest. Its mate walked up to the nest with chick number one following, and took charge of all three

Snow geese fly above the tidal flats of Hudson Bay ▶

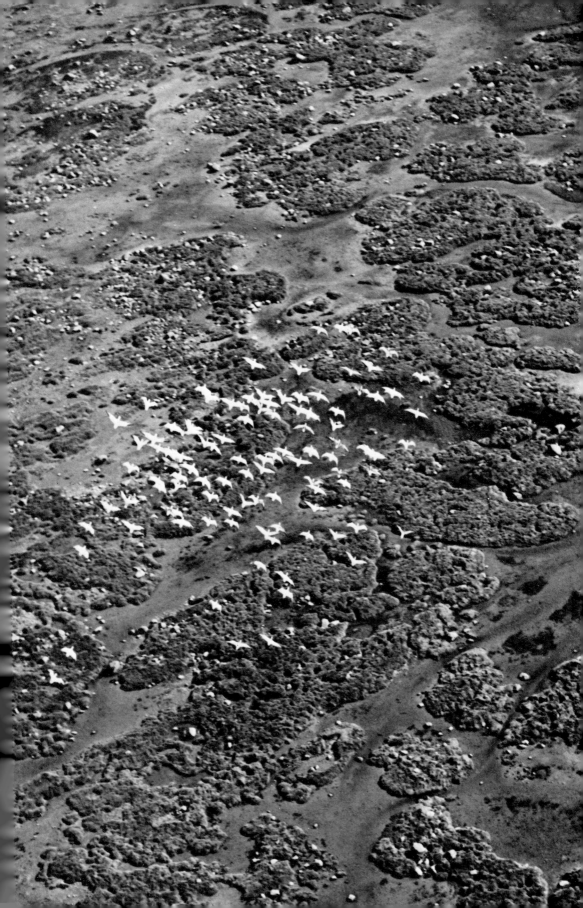

The imprinted goslings, between 3 and 8 days old, having a rest

The goslings stay close to Des as he films a golden plover on the nest

The incubating golden plover is well camouflaged

The 'Creeps' catch mosquitoes off Des' down jacket and Les' jeans

At 5 weeks a snow goose is almost ready to fly

The goslings 'help' Jen with the laundry

Chewing on a tent rope

The 'Creeps' having fun on the tundra lakes

chicks and the last hatching egg. The dry chicks made short excursions within a radius of a yard or two of the nest, balancing on their long-toed feet and pecking inquisitively at everything, but the unconcerned parent showed no further interest in them. We had no way of knowing which was the male and which the female, nor were we able to watch other pairs of semipalmated sandpipers closely enough to know if this changeover of responsibilities follows a normal behaviour pattern.

The interesting pectoral sandpiper also nests at the McConnell River but is uncommon. The most unusual thing about this larger species is the fantastic display of the male early in the nesting season. Flying several feet above the ground, he makes strange booming calls while blowing out his throat and chest like a balloon. It really is quite a remarkable performance and one that we witnessed only two or three times, but never managed to capture on film. We found only two nests of pectoral sandpipers and both were destroyed before the eggs could hatch.

There is no space to mention in detail all the fascinating nesting birds we encountered on the tundra, but one which we shall never forget is the beautiful golden plover. Once abundant, these handsome birds were almost wiped out by hunters near the turn of the century. They are now protected and their numbers are beginning to recover, even though the golden plovers are international air travellers and therefore need protection in each country they visit. We had first come across them on the Hawaiian island of Oahu, when they were in non-breeding plumage. Those particular birds belong to the population nesting in Alaska, whereas the golden plovers nesting in the Canadian north fly all the way to southern South America during the northern winters.

Although we had seen a few golden plovers at the McConnell River, we had almost given up hope of finding their nest when John Harwood returned to camp one evening with the news that he had seen a pair behaving as if they had a nest nearby. Les was eventually able to find this nest, the first ever discovered at the McConnell. The golden plovers are so prettily marked with black, white and yellow that one would expect to be able to spot a sitting bird with ease, for they nest on the drier parts of the tundra where only lichens and the tiniest plants grow. But with the head tucked low to compress the conspicuous white markings on the side of the head and neck, the bird blends in perfectly with its surroundings. At first we used a blind to photograph the golden plover on the nest but soon found this to be unnecessary, for we could work in the open only a few feet away from the sitting bird without it leaving the nest. The only problem we had with this pair of birds was that their eggs took a very long time to hatch. We particularly wanted to photograph their tiny chicks so we had to do a lot of walking as the nest was across the river and three miles from camp. Every second day one of us would check on the eggs until finally one egg showed the first signs of starring. With many species of birds, this might indicate that the chicks would hatch the following day, but three days passed before the little yellowish plover chicks emerged. Every day we had to make the trip across the river to their nest just in case, and on the day they finally appeared drizzling rain was falling. Not the best weather for photography.

One midsummer evening while planning the next day's filming, I remarked to the others that on no other trip in my twenty years of wildlife filming had past experience mattered so much as on the tundra. The weather was unreliable, and everything happened so quickly during the short summer that it would be very easy to miss filming some important natural history event which was essential to the overall snow goose story. For although our main project was to make the one-hour television special, *The Incredible Flight of the Snow Geese*, for Survival Anglia Limited, we also wanted to film as much variety of wildlife as possible for their series of half-hour SURVIVAL programmes. In the United States these programmes are called 'THE WORLD OF SURVIVAL'. The boxes of exposed film mounted rapidly and we could only hope that none of the cameras had developed a hidden fault. There was no way of sending any of the exposed film out for processing and checking until we ourselves left the tundra at the end of the summer. However, all went well, and four half-hour films were finally made: *Land of the Loon, Arctic Summer, On the Trail of the Snow Goose* and *Central Flyway*. Each shows something of the wildlife filmed during the two-year period 1971–72.

6 The Land of the Loons

The Creeps grew at an incredible rate and by the time they were three weeks old they weighed almost three pounds each, and feathers were beginning to appear on their wings and tails. They were no longer cute little balls of fluff, but ungainly young geese. The colouring of the snows had faded to a fairly uniform pale grey, while the blues were a dark grey. Now we could leave them completely free to wander round the camp at night as well as in the daytime. At this age their main occupation was eating; the rest of their time was spent sleeping, swimming and preening, or helping us with whatever chores we happened to be doing. Most of all they liked to be with us: puddling with muddied bills in a bucket of soapy water on laundry day and tugging at the clean clothes; chewing on rubber hip boots left outside the tents to dry; repeatedly untying our boot laces in camp; and chewing the tent guys until the nylon ropes were shredded by their strong bills. On many mornings this noisy tugging on the tent ropes woke us up, but we could hardly feel annoyed when we opened a bleary eye to see a Creep's shadow clearly outlined on the canvas tent wall by the early morning sun around 4 a.m. They were compulsive chewers, but not very selective ones – they sometimes even chewed on us. Although their motive was undoubtedly affection rather than malice, they often pinched our flesh so hard with their bills that even through our clothing, dark bruises resulted.

On sunny evenings in July and August, when we arrived back at camp hot and tired after a long day of filming, we occasionally braved the chilly river water and the clouds of mosquitoes for a short but refreshing dip. The Creeps loved to follow their 'parents' into the river, swimming in a cluster around us and pulling at our hair or anything else they could find to chew. They were so at home in the cold water and enjoyed it so much, that they must have been puzzled by our eagerness to leave it after only a brief swim.

Whenever we left camp the Creeps would be sure to follow, but there were times when we knew they might interfere with our filming and we had to resort to trickery in order to leave them behind. While we headed away from camp, one person stayed to keep the geese out of sight behind the tents. Once they had settled down to preen and rest, the last person sneaked out of camp, but the wily geese were not always fooled. Like most children, they hated to miss out on anything interesting.

To conserve our energies we sometimes left heavy equipment carefully stacked in the blinds overnight if we expected to film in the same area the following day. This worked well until one night in July when, after a clear sunny day, a severe storm hit without warning; it was so violent that in spite of all our precautions, some of the boxes in the main tent at camp became wet. During the second night the storm abated and the following morning we set out fearing the worst. On reaching the first blind where we had been filming nesting Arctic loons and Arctic terns, it was a relief to find the tent still standing. But when we looked inside we were shocked to see that the large camera case, containing a Nagra sound recorder and still cameras, was sitting in water several inches deep. After pouring about three cupfuls of water out of the recorder, I immediately removed the twelve 'D' cell batteries which were already leaking acid badly. One 135 mm Minolta lens was so full of water trapped between the elements that it resembled a kaleidoscope with tiny pieces of floating matter creating ever-changing patterns as the lens was rotated. Depressed, we carried the drowned equipment back to camp, where we dried it out as well as we could, using paper towels. The Nagra was then placed upside down on a high shelf in the cabin and left to dry out thoroughly. Amazingly, when two weeks later we put in new batteries, this rugged sound recorder still worked.

While we stayed in camp, Les hiked over three miles to a blind where we had left our main movie camera, a 16 mm Arriflex with several lenses, wrapped in a sleeping bag and placed inside a waterproof duffel bag. Luckily very little water had entered this tent so that the camera was quite dry and in perfect condition. We then realized that the water in the first blind was not merely the result of the heavy rains. The strong winds must have driven waves across the lake and over the bank to flood the tent and our equipment. After this experience we never left cameras in any of the blinds overnight unless everything was first sealed in large plastic garbage bags for extra protection and then placed off the floor on rocks. The waterlogged still camera gear was useless until it was repaired many months later, but as we always carry extra equipment we did not miss any photographic opportunities during the rest of the Arctic summer.

Later in July and in August, we often had sudden but short rainstorms in the afternoons. Though the sun would be shining brightly over most of the tundra, patches of threatening black clouds would be visible overhead. After the passing of one of these storms, we were sometimes treated to the beautiful sight of a complete rainbow. On the flat, open tundra the multi-coloured arch could be viewed in its entire splendour. Our 'pot of gold' came from the spiritual lift such perfect beauty gave us.

By mid-July the sea ice on Hudson Bay had broken up sufficiently to enable the biologists to travel to Eskimo Point in their freight canoe, but this was a trip that could only be undertaken when weather conditions were ideal. At low tide the water receded about five miles into the bay, exposing boulder-strewn mud flats which were difficult to cross on foot. All canoe journeys therefore had to be timed so that they began and ended on a high tide at the McConnell, where the freight canoe was left upturned on a high point in the delta when not in use. Powered by an outboard motor and dodging floating chunks of ice, the heavy, twenty-two-foot-long canoe could make the thirty-mile trip to Eskimo Point in about three hours. Throughout the summer Charlie maintained radio contact with Father Ducharme at the Catholic Mission at Eskimo Point, or with the branch manager of the Hudson's Bay Company there. Should one of our party have been injured or fallen seriously ill, a seaplane would have attempted the hazardous landing on the swift flowing, boulder-strewn river, but fortunately we had no such emergency.

Now that the freight canoe could be used, we were in partial contact with the outside world. Once during June two of the biologists, Dave Ankney and Larry Patterson, had walked to Eskimo Point, a gruelling thirty-mile hike in hip waders. Their main purpose had been to take some specimens to store in a deep freeze in Eskimo Point, but they also brought back mail for everyone in camp. On two or three other occasions, small parties of Eskimos called at camp with mail, on their way up the McConnell River to hunt caribou. The Eskimos always had an enormous appetite for bread and jam and one could say they drank their sugar with a little tea added! Cheerful people, they were formerly inland Eskimos who had depended almost entirely on the caribou for their livelihood; now they had been resettled on the coast where they could hunt seals, walrus, beluga whales and fish at different times in the year. Although we did not really expect visitors in the remote camp, any letters that had been written were always kept ready in a plastic bag 'just in case'. On at least one occasion this proved useful. We were far out on the tundra when we heard the unfamiliar thump – thump – thump of a helicopter and watched in surprise as it landed close to the cabin. The pilot was simply being sociable and had called in for a cup of coffee. He left half an hour later, taking everyone's mail for posting.

By 20 July most of the adult geese had moulted their flight feathers. This rendered them flightless for almost a month, until new feathers grew. A further drain on the adult's strength, this annual moult is timed so that they are still flying and able to protect their goslings during the most vulnerable period when the chicks are tiny. By mid-August, when the young are learning to fly, they must be airborne again.

Taking advantage of this flightless period, Charlie and his students conducted a series of banding drives. Walking for miles across the tundra and keeping in touch by walkie-talkie radios, they slowly rounded up small family parties and groups of geese. By the time they reached the previously constructed holding pens there were several hundred geese grouped together, mainly snows and blues, with a sprinkling of Canada geese. It took several hours of hard work before the geese were safely in the pens. Holding the birds for as short a time as possible, the scientists keep detailed records as each one is sexed, banded with a numbered metal ring around one leg, and released. By now the goslings are large enough to wear the permanent leg bands. The temporary, numbered web tags, which had already been placed on newly-hatched goslings, are important for research relating to brood size and breeding success, and enable accurate information to be kept of parentage, hatching dates, and so on. In this day and age, computers have to be used to help analyse the mass of data accumulated during detailed studies of geese and other birds.

Details of all bird banding carried out in North America are sent to the U.S. Fish and Wildlife Service. Any hunter who shoots a banded goose, or any person who finds a dead, banded bird, follows the instructions on the small metal band and mails it to the Migratory Bird Population Station in Maryland, together with details of where and when the bird was found. In return, details of the bird's history are then sent to both the bander and to the person who recovered the band. Through keeping such detailed records, biologists have been able to work out over the years the migration routes used by the geese from the various nesting colonies. These and other studies benefit the geese in the long run, for armed with more knowledge about them, biologists are better able to plan for their needs.

Some of the geese at the McConnell River were fitted with coloured neck collars so that their movements could be more closely studied among the thousands of snow geese on migration and on the nesting grounds in subsequent years. Such studies made by J. Paul Prevett from the University of Western Ontario, between 1968 and 1970, followed the family life of lesser snow and blue geese breeding at the McConnell River. Studying the geese on the nesting grounds and at stopover points along the migration route, Paul and his wife Lynda spent hundreds of hours peering through a telescope in order to identify individual birds by the coded letters marked on their neck collars. They discovered that every family group remained together on the long southerly migration each fall. Paul observed that hunting pressure often splits up a family group, but later the family reunites. A classic example of this occurred in Texas on the rice prairies west of Houston, where a large proportion of the snow geese spent the winter. One Friday, Paul noticed that a particular family of geese was no longer together, and later in the day discovered the missing juvenile 100 miles to the west of its family. Then, amazingly, on the Monday all the geese belonging to this family were seen together again. This is an astonishing example of how snow geese can identify the calls of members of their own family from the bedlam of calls made by thousands of other geese; it also shows how a young goose, once separated, goes from one flock to another searching for its missing family. In addition it reveals why young geese may so easily be decoyed down to waiting hunters.

In late July we had additions to our growing family of orphans. First, Charlie and Lee saved a hatching Arctic loon from a parasitic jaeger. Lee carried the loon egg around inside her shirt until the bird emerged, none the worse for its adventure. When dry, the chick was a bundle of light grey fluff with, for the first hour or two, a large bare orange patch on its belly. This bulging yolk sac gradually shrank into its body until the underside appeared to be completely covered with down. Once again, food was the problem. There were plenty of tiny fish in the stream beside the camp, but catching them was another matter. We could never compete as fish catchers with the true loon parents, though by building small rock dams we gradually became more efficient, and with patience we were soon able to feed up to a hundred little fish a day to the 'lunatic', as he had become known. Strips of fish cut from large grayling were dipped in Miracle dog food crumbles for additional variety and nourishment. If it survived, we knew that the loon would have to live mainly on frozen fish and Miracle when we travelled south following the snow geese. Sadly, after we had had it for six weeks, the loon died just as its feathers were almost fully grown. There seemed to be no obvious cause for its death, unless this were due to a dietary deficiency which would be most seriously felt during the feathering out period. Later we learned from people who had tried to rear loons that they are notoriously difficult to keep going through this period of rapid growth.

We were also landed with seven orphaned greater scaup ducklings; they were even more difficult to feed than the baby loon. This was partly because there were so many of them, and also because, unlike loons, they normally feed themselves. Had they been with their real mother, the little ducklings would have swum in the tundra ponds and lakes, dabbling and diving for daphnia, fairy shrimp, tiny fish, small snails, insect larvae and any small animal they could catch. We couldn't spend the day swimming with them, but we netted whatever pond life we could find. Dropped into a large basin of water with the ducklings, everything we offered them was gobbled up in no time. We tried to get them to eat the high protein dog food, for we knew we hadn't a hope of raising them on their natural foods alone. After several days, two or three began to take a little of the Miracle when we dropped tiny pieces into a dish of shallow water. Outdoors we let them swim in small pools whenever we had time to watch over them. They loved to dive and dabble at the muddy bottom where we assumed they found food. On the tundra they pulled at grasses and seeds and were expert at taking mosquitoes off our clothing. Sad to say, almost every day we had one less duckling, until by the end of a week only two remained. These two had

Snow goose leading goslings to the McConnell River

Off for a day's filming

The grace and power of a flying snow goose
One thousandth of a second captures a blue goose tumbling

After a bath a snow goose flaps its wings

gained weight rapidly once they began eating the Miracle and they grew apace. Luckily they turned out to be a pair and were later given a good home at the United States Fish and Wildlife Service 'Northern Prairie Research Center', near Jamestown, North Dakota, where they will be used for breeding studies.

Some Arctic terns probably travel farther each year on their migrations than any other species of bird. After breeding in the far north, most cross the North Atlantic to Europe, fly south off the west coast of Africa and cross to South America and the Antarctic, before returning to their tundra nesting grounds. So during the northern hemisphere winters they fish in waters to the south of southern Africa and South America. They are identical in appearance to the Antarctic tern, but when the northern breeders are in southern latitudes they are in non-breeding plumage, sporting more white on their heads and darker beaks than the southern species, which at that time is in full breeding plumage. So although both species may be seen in the same areas during the southern summer, they can usually be identified by the seasonal differences of their plumage. When very young, the chicks of these two species are also easily told apart by their quite different markings.

At first we did not think the Arctic terns nesting at the McConnell River were at all friendly. Whenever we walked within forty yards of one of their nests the pair flew above us, screaming loudly and diving aggressively at our heads. Hastily pulling up our down jacket hoods we could still feel the whack of the beak each time a tern hit us, and we found they made tiny holes in the fabric. Ornithologists say that the way to tell an Arctic tern from other species of terns is this: all terns will dive at intruders in order to protect their nests but if the bird draws blood, then it must be an Arctic tern! When we set our cameras on tripods to film the terns they dived at the cameras, striking lenses with their sharp bills before zooming high to dive again. One hit my hand while I was filming and it most certainly drew blood.

We found one pair of Arctic terns nesting on a small island in a lake not far from camp, and at first whenever we went near their nest they attacked just as aggressively as any other pair. But after a few minutes of noisy dive-bombing, first one, and then the other, would actually settle on our heads. Les often wore a woollen cap and this seemed to fascinate the terns, for they tugged at it repeatedly. Otherwise they seemed to think that our heads were just fine resting places on which to regain their energies before launching a fresh diving attack. We were never

◀ Snow geese flying above Sand Lake Refuge

able to think of any other reason for their behaviour, and although we spent a considerable time near several other tern nests, none of the other birds behaved in this way.

The tern chicks leave the nest soon after they are dry and we watched one tiny chick swim from its nest island in a lake to a large flat rock patterned with patches of bright orange lichen. Here the parent landed repeatedly with tiny fish to feed the speckled youngster, but on one occasion brought a fish that was too large for the chick to swallow. After trying unsuccessfully to feed it to the chick, the adult tern finally ate the fish. When it had been an hour or more on the rock the chick, encouraged by the adult, eventually swam back to the nest island where the other parent was incubating the remaining pipped egg. As the tern chicks grow they are fed larger fish, keeping both parents busy satisfying their tremendous appetites. Only once did we see Arctic terns offer anything but fish to their young, and on that occasion a tiny chick hungrily devoured a crane fly.

On a small island well out in the lake we found an Arctic loon nest. It was simply a shallow depression on the damp ground, lined with a few pieces of vegetation and linked to the water eighteen inches away by a muddy slide. Loons appear to have incredibly short legs which are situated far back on the body; in fact, the upper part of each leg is hidden within the body. Because of this the loons are very awkward on land. Unable to walk, they shuffle a few paces in a semi-erect position or just slide along on their bellies when out of the water. For this reason their nests are always within a few feet of the water, and once in the lake they are in their element, for they are great swimmers and divers. In Europe the common name for a loon is 'diver'. Loons are heavy-bodied birds with rather short, narrow wings. In relation to their overall weight, they have the smallest wing surface of any species of bird and therefore fly with very rapid wing beats.

We were extremely pleased to find an Arctic loon nest to film so close to camp, but there was a problem. Their small nesting island was surrounded by water two feet deep, and there were no other islands suitable for a blind within fifty yards of the front of the nest. Obviously there was only one thing to do: make our own island. With the help of the Argo, Les and I ferried many loads of rocks out into the lake and soon had an island, just large enough to accommodate a blind, only twenty feet from the Arctic loon nest.

The loons settled down well and our comings and goings did not upset them at all, as they were used to seeing and hearing humans a few hundred yards away on the camp ridge. Because of the difficulty of securing a tent blind on our tiny rock island, we used instead a large plywood box that was just high enough to sit in comfortably.

The loons nested a little later than most other birds on the tundra so we were able to spend more time watching and filming both the Arctic and red-throated species. We quickly became loon fans! It is impossible to imagine a more beautiful bird than the Arctic loon in its summer plumage, although with its pale grey head and black and white body and wings, one can hardly call it colourful. The sexes are identical in appearance. We spent many happy hours watching the Arctic loons from close quarters as they returned to the nest, rolled the two large brown, blotched eggs, and settled down to incubate. Watching alertly with bright red eyes, the sitting loon flattened on the nest, with neck stretched low to the ground, if there was any sign of danger. Whenever anyone approached the blind, the sitting bird slid inconspicuously into the water and submerged with scarcely a ripple as it swam well away from the nest underwater and therefore out of sight. By doing this, the crafty loons never disclose the location of their nests. However, during the summer, we found that whenever a pair of loons took possession of a particular lake, their nest could be found somewhere around the shore or on an island well out in the water. Each pair would defend their territory from all other loons, but on the tundra there are plenty of lakes for each pair to possess one privately. When we approached their nest the first few times, one of the Arctic loons splashed about on the water thirty feet away, flapping helplessly as if injured as it attempted to draw our attention away from the nest.

We were not the only ones to be dive-bombed by the Arctic terns. Although nesting on adjoining islands, the terns would not tolerate an Arctic loon swimming near their island and would dive on it unmercifully, causing the loon to hold its head low in the water or even crashdive for safety. When their two light grey, downy chicks hatched a day apart in early August, we felt privileged to watch the loons' intimate and affectionate family life and to see the way the adult loon removed the eggshell from the nest after the second chick hatched. Without leaving the water, one adult reached into the nest and lifted a large piece of eggshell in its bill. Swimming several yards away, the loon dipped the shell into the lake a few times before releasing it, to make sure it would sink out of sight. One adult then returned to the nest to brood both chicks. Later we were to see a red-throated loon dispose of an eggshell in exactly the same way.

For the first two days the Arctic loons kept their chicks at the nest. Occasionally one youngster became adventurous and tried a little swim on its own, but never ventured more than a few feet from the nest. One adult remained on the nest with the chicks sometimes completely hidden underneath, but more often their heads peered out from under a wing, or they came out and snuggled alongside the sitting parent. The other adult was kept busy fishing in the lake and about every fifteen minutes swam right up to the nest with a two-inch-long fish in its bill. If the chicks were hidden beneath the brooding adult, its mate made low calls as it neared the nest and a hungry chick would quickly appear. Without leaving the water, the loon stretched towards the nest and offered the fish to one of the chicks, which usually swallowed it immediately. If the chick dropped the fish the brooding adult then picked it up and offered it again and again until it was eaten. Once we counted as many as eight little fish being brought to the nest within one hour, but only on one or two rare occasions, when both chicks were too full to eat any more, did the adult itself finally have to swallow the fish. Between feeding sessions the two chicks constantly pecked at each other's heads, but the sitting adult usually tried to keep them separated by brooding one under each wing. During the first two days when we were able to watch the chicks closely, there was no competition between them for food. However, this was before their appetites became so great that they would strain the parents' food gathering abilities, and then if both chicks were hungry the strongest one would grab the most food. In fact, although loons lay two eggs, they frequently raise only one chick to flying age.

After the second day the loons left their nest for good, although they sometimes returned to it at night to brood the young. Before leaving the tundra at the end of August, we often saw this loon family far out on their lake with the two young being fed frequently as they swam along behind the parents. On another lake we sometimes visited, a single Arctic loon chick was constantly attended by one adult while the mate was off feeding elsewhere. Pausing to watch more closely, we were fascinated to see that the half-grown youngster was being fed entirely on fairy shrimps, the adult catching and feeding the chick as many as fifteen of these little creatures in one minute! Dipping its head just beneath the surface, the adult soon reappeared with a fairy shrimp held in the tip of its bill. This was quickly given to the waiting chick and the process repeated. Presumably the chick was sometimes brought a fish from another lake, but we were never present when the mate returned.

The slightly smaller and more slender red-throated loons do not have such a beautifully patterned body as the Arctic loons, but their rusty-red throat is very distinctive when seen at close range. While every Arctic loon nest that we found was on an island, each of the red-throated nests we located was on the bank of a lake or river, but always within a foot or two of the water. The young red-throated loons were a much darker grey than their Arctic cousins, and although they were also fed on fish, the feeding method was totally different. Taking it in turns to remain at the nest or elsewhere on the bank with the young chicks, one adult flew off towards nearby Hudson Bay, returning an hour or more later carrying a single fish in its bill. But what a fish! They were often sand-lances five or six inches in length, and longer than the chicks that were supposed to eat them, yet somehow the tiny loons managed to swallow these fish whole. They were always taken head first, but the tail often protruded from the chick's bill for half an hour before room could be found for it inside!

In contrast to the Arctic loon chicks, these were always fed in the water, even when tiny. After a spectacular belly landing with landing gear up, the fish-carrying adult swam straight towards the nest making moaning sounds to alert the chicks that food was on its way. One or both chicks then flopped into the water to meet the returning adult and take the large fish. This was quite a performance. The adult kept a good grip on the slippery fish until the chick began to swallow it, but once the youngster had the full weight of the fish to support, it had great difficulty in keeping its own head above water. Tiny feet paddled frantically as the chick tried desperately to keep its head up, until at last the long fish miraculously disappeared and the chick's equilibrium was restored. As soon as the chick had swallowed the fish, the baby-sitting red-throat took off for its turn to fish in Hudson Bay.

A loon needs a long water runway for take-off so that it can run into the wind flapping, its feet splashing on the surface of the water, until it gets up enough speed to lift into the air. It is physically impossible for loons to take off from dry land. When landing, again always on water, loons do not put their feet down to act as brakes on impact as waterfowl do, but leave them trailing behind and land on their bellies, occasionally bouncing in the air spectacularly if the water is rough and they hit a wave crest.

Once when we were in a blind filming red-throated loons and their young, a third adult of the same species landed on their pond. Immediately both resident loons ran across the surface of the pond towards the intruder, flapping aggressively and holding their bodies erect so that only feet and tails touched the surface. The pair actually took off and flew after the interloper, chasing it for some distance before returning to belly-land on their own pond again. They then swam towards each other with necks stretched low on the water, opening their bills as they uttered their raucous cries. The varied and haunting calls of both these species of loons will remain with us as one of our most vivid memories of life on the tundra. Lying in our sleeping bags at night, we heard them make all kinds of moaning noises, and other calls resembling a baby crying, or a zebra braying in the distance. All five species of loons are notoriously vocal, especially during the breeding season and at night. Their mournful, far-carrying cries range from yodels, howls and wails, to unreal maniacal laughter.

On 19 August, when our red-throated loon chicks were just a week old, there was a sudden violent storm bringing with it bitterly cold temperatures. Afterwards, we could find only one loon chick. Evidently the smaller one succumbed to the cold and lack of food during the rough weather. We found that the young of some other species of birds, such as Lapland longspurs and savannah sparrows, had also died during this late season storm. These must have been birds which had lost their first clutch and re-nested, for the chicks that we found dead in the nests were still very tiny.

During the summer we saw caribou quite often, but only single males or small groups of three or four animals. They usually trotted off with their distinctive high-stepping gait whenever we came within fifty yards. However, as we emerged from our tent early in the morning on 11 August, we noticed twenty or thirty caribou dotted about the tundra feeding within a half-mile radius of camp. Using a high-powered telescope to check in all directions, Les counted a total of 412 caribou scattered across the tundra. This was by far the greatest number that we noted on any one day. Evidently the biting insects were more bothersome inland and the caribou had moved towards the coastal flats to escape.

Apart from the caribou we did not see very many mammals during our summer on the tundra; two polar bears; about half a dozen weasels; one lone wolf; one Arctic fox and the Arctic hare mentioned earlier. We also saw several of the large reddish-brown Arctic ground squirrels, called shick-shicks in this part of Canada. On 1 July near camp the first one had shown itself above ground after its long winter hibernation. Moving along on all fours close to the ground, it nipped off a mouthful of grass and sat upright to eat, with front feet acting like little hands to stuff the protruding grass stems into its mouth. This ground squirrel also relished the blooms of the mountain avens growing along the dry ridge near camp. After a few days it moved off and we did not see it again all summer, although quite a number of these charming little rodents lived along the sides of an esker a few miles south of camp.

7 Time to Head South

By the first week in August the Creeps were covered with new feathers, showing traces of down only on the head, neck and lower back. The first feathers to push through the down had been on the tail and shoulders, followed a little later by others on the belly and breast. Once their feathers appeared, the goslings spent a great deal of time preening, paying particular attention to the flight feathers of both wings. As they were the last feathers to grow they gave the wings a very ragged and unwieldy appearance during the growth period. The snows had mainly white bodies, with wing, head and neck feathers a light grey washed with fawn. The blues had light bellies and light lower backs, with the remainder of their plumage various shades of dark brownish-grey. The Creeps' voices had deepened and they attempted, with limited success, to make adult goose noises. Sometimes they sounded more like a goose with hiccoughs – the equivalent of a human teenager with a breaking voice. Their wings were now exercised regularly as the Creeps ran and flapped across the tundra, and if we ran ahead this encouraged them to try even harder to become airborne. Clumsy at first, they gradually gained control as their feathers and wing muscles strengthened. At just six weeks old they at last made their first short flights while flapping as they ran into wind. Each day they flew a little farther and a little higher, to begin with always in a straight line. But less than a week after the first had lifted a few feet off the ground they were making circuits of the camp, although their landings were sometimes a little bumpy.

Soon the Creeps were flying out on to the tundra to pay surprise visits to the loon blinds where we were filming. They could not see us but somehow knew when we were inside. After greeting us with a noisy landing they wandered about feeding before settling down alongside the tent to sleep until we had finished for the day. How we envied them as we carried our heavy packs while they flew effortlessly on ahead to wait for us in camp. Sometimes they flew far out across the tundra, mingling with the wild geese before they disappeared from sight in the larger flock. Much to our relief, they would be back after a few hours but we couldn't help wondering if it was affection for their foster parents that drew them back – or a liking for Miracle dog food! Now that the young geese were flying they were no longer goslings but had officially become juveniles. At six months they would be called immatures and at about twelve months would acquire adult plumage.

The Canada goose and the sandhill crane were unable to fly until a week after the young snow and blue geese. Perhaps this caused the beginning of the close friendship that developed between these two and lasted almost a year, in fact as long as they were together. They made a strange pair: Fred tall, long-legged and graceful, and his squat friend waddling alongside on short legs. When they too could fly, they once more went everywhere with the rest of the Creeps. At first the colouring of the Canada goose had been very similar to that of the snow goslings, but as feathers displaced the down on his head and neck at six weeks the white chinstrap showed clearly among the black feathers and there was no mistaking his identity. Our Canada goose was always much more gentle and friendly than the snows and blues. He did not even mind being picked up and stood on one's lap, revelling in the attention. Although Fred never liked being lifted up or even touched, he too became very spoiled and was often jealous of sharing our attention with the other 'children'.

As August progressed, the nights became quite dark for an hour or two. For the first time that summer, on clear nights we could see the lights of Eskimo Point twinkling to the north of camp; they appeared to be much closer than thirty miles away. Each night now we watched for the northern lights, or aurora borealis, and were treated to several spectacular displays of constantly changing bands of greenish-yellow light rippling across the sky. Our most vivid memory is of one particularly impressive effect created in the night sky. It was as if we were gazing upwards inside a gigantic cathedral, with vertical streaks of light making ever-changing patterns on the inner walls of the huge dome overhead. Tired as we were, we found it difficult to turn in for the night with such a glorious display going on overhead.

There had been many spectacular sunsets during the summer, but the ones we remember best occurred on calm evenings in late August. Then we liked nothing better than to watch from the shore of a tundra lake or pond as both the clouds and the still waters caught fire, their shades of brilliant red and orange contrasting with the blue of evening. All around us peace and quiet reigned.

During the first two weeks of July we had rarely seen anything of the wild geese except as white dots in the far distance on the coastal flats. Later they seemed to scatter all over the tundra, and sometimes we rose in the morning to find several families feeding within a hundred yards of camp. However, they usually went farther off as soon as they saw people moving about the camp area. Walking across the tundra, we never came within a quarter of a mile of the wild geese, for although they were flightless they could see us approaching from far off and moved well out of our path.

By mid-August most families of blue and snow geese were flying. At first each family flight was very short, but we noticed that they almost always flew southwards on these short hops. By the end of August a big percentage of the McConnell River geese had begun their migration, following the west shore of Hudson Bay southwards in leisurely stages as they fed and regained the strength they had lost during the breeding season and moult. We hated the thought of leaving the peaceful tundra, but we too had to be gone before the first winter storm descended on the area.

In previous years the biologists had used their big freight canoe to ferry people and gear to Eskimo Point at the end of the summer. We weren't too happy at the thought of making the canoe trip with salt spray blowing over the camera equipment and exposed film – the result of three and a half months of extremely hard work. So we discussed one alternative with Charlie, who thought it might be possible to fly out if we could find an area dry enough for an airstrip near camp. It would save the tedious business of moving equipment over the three miles of rough terrain to the coast, where the canoe was kept.

One day in mid-August Charlie found a possible place for a landing strip a mile up river from camp. After clearing the area we marked the improvised airstrip with orange flags. Charlie then radioed Eskimo Point, requesting a plane from Churchill on or after 26 August.

We got no answer on the radio for several days, then suddenly word came through at 7 p.m. that the plane would come early next morning. The following hours were very busy ones, everybody working far into the night collapsing tents and packing up equipment. When the radio message reached us, I had the Argo in pieces making essential repairs, but within an hour it was ferrying heavy loads to the new airstrip. There everything was cached safely under heavy plastic sheeting so that loading could begin as soon as the plane landed – provided the pilot decided after a closer inspection that our airstrip was safe. We knew that not everything would fit into the first plane load, so we could keep one tent up for sleeping. For the next two days the weather was fine, but although we watched and waited there was no sign of a plane. We carried on filming close to camp, and sighted a polar bear half a mile down river. In fact, the bear had actually passed through the camp the previous night, but had not caused any trouble. All summer we had been disappointed at having seen only one other polar bear. He had appeared in July and spent two days feeding on an old caribou carcass near the coast. Whenever we tried to approach he moved away, so we had no luck filming him. Perhaps this time we would be more fortunate. The polar bear moved about slowly among a maze of channels and low islands in the river delta, and as we cautiously approached on foot he waded out to a large rock and stretched out in the sun to relax. As he showed no objection to our presence fifty yards away we spent some time photographing him, with a group of snow geese feeding on the bank in the background. But when we tried edging a little closer the bear decided he had had enough of our company. Slowly he sat up and stretched, before wading across the river and ambling off towards the coast.

As we stood watching, we suddenly became aware of the drone of an approaching plane; anxiously we watched as the Twin Otter circled the tiny landing strip, then breathed a sigh of relief when it touched down. However, when we reached the plane we found that its wheels had bogged down just off the edge of the flagged area. It seemed that heavy rain four days earlier had softened the spongy tundra more than we had realized. After much digging, boards were fitted beneath the wheels and by using full throttle the pilot was finally able to coax the plane free, and taxi on to sheets of plywood before loading began.

Six of Charlie's team flew out on the first trip and we watched tensely as the plane roared across the tundra. Would the take-off go smoothly? Having checked out the airstrip thoroughly on foot, the pilot had calculated his load carefully and used the full length of the strip before lifting the plane into the air. We were relieved to see the Twin Otter safely airborne, but now began to wonder whether it really would return early the following morning as promised – perhaps the pilot had had more than enough of our McConnell River 'Airport'.

Over a period of two days it took four trips to fly everything the one hundred and seventy miles to Churchill. Unfortunately the Argo could not be fitted into the Twin Otter, so I stayed behind with Les and Charlie to drive it to Eskimo Point. By now the little vehicle was showing the effects of all the hard work it had done during the summer. Twice spare parts had been sent in to keep it running, but it had been out of action for long periods. By driving carefully, with the two passengers walking over the rougher sections, we managed to make it to Eskimo Point. We reached the small settlement to find that the yearly supply ship had just arrived to restock the Hudson's Bay Company store with supplies for the next twelve months. This was a stroke of luck for us, as we were able to make arrangements for the Argo to be sent to Churchill on board the ship. We then flew there ourselves on the first available plane, taking a farewell look at the McConnell River as we passed high overhead on our way south.

Meanwhile Jen and Lee had flown out on the third plane trip from the McConnell River to Churchill, travelling with the Creeps. For the first time in their lives the birds were confined and, not surprisingly, they disliked being shut up in their boxes of plywood and wire netting. We had been concerned that they might fly out on the tundra and disappear from sight when we wanted to load them, but we needn't have worried. Like the pet dog in any family when its owners are packing for a trip, the birds knew that something was going on and stayed very close to us as the tents were taken down.

From the air we saw many groups of snow geese near the shore of Hudson Bay, and their numbers increased steadily as we flew southwards. Their migration had begun, and so had ours, for we would follow the wild geese all the way to their wintering grounds in Texas and Louisiana, 2,500 miles to the south.

After three and a half months on the tundra, returning to the relative civilization of Churchill had its surprises for us. Driving a car again was no problem, but even at the sedate speed of twenty-five miles an hour we all felt uneasy at the great sense of speed this produced after months of slow travel on foot, or in the Argo. After handling no money all summer, paying cash for anything required great concentration, especially when we found a mixture of American, Canadian and Australian currency had accumulated in our wallets! Presented with a menu at one of the two local hotels, we were confused by the choice of so many good things to eat, but fresh eggs proved to be the most popular item. Later, in the supermarket, our eyes boggled at the sight of delicious fresh fruit and vegetables, but our first hot shower was perhaps the greatest luxury that civilization could offer us.

For the first few days in Churchill we camped with our birds in a large Lambair hangar, well away from the main airport buildings. Outside on the grass we erected a wire pen where they could exercise all day. Not knowing how they would react to a strange area we were reluctant to allow the Creeps to fly at the airport, where there were many hazards such as high tension wires, buildings and even stunted trees. The geese might also have been a hazard themselves, especially if they flew near the runway when a plane was taking off or landing. They turned out to be very adaptable and settled down quickly to their restricted routine, but Fred was unhappy and for a few days ate very little. More highly-strung than the geese, we later found that he was always far more upset by travel and major changes in his routine. Our birds created much interest at the airport and many people came from the town to visit them, for although they often saw geese flying overhead, they rarely had a chance to see them at close range and completely tame.

The airport was not an ideal place for the birds, but it was a great help to be able to keep them there until we found somewhere more suitable. Charlie introduced us to Father Paradis, who kindly suggested that we might like to camp beside a small lake fifteen miles east of town where the Catholic Mission owned a small cabin. Gratefully we accepted his offer, setting up our tents behind the cabin and building a large wire pen straddling the lakeshore so that the Creeps could once again enjoy swimming and bathing. After their carefree life on the tundra we hated to confine them, but the goose-hunting season had started around Churchill on 1 September, and we feared for the Creeps' safety. As we followed the wild geese southwards, every time we let them fly free we would be faced with the risk of our tame geese being shot. Wherever we followed the migrating wild geese, the goose-hunting season would be open, from September in northern Manitoba until mid-January down in Texas. We had decided that our birds should be allowed to fly free each day, partly because they enjoyed it so much but also to keep them fit.

Because of the cabins built around the lake where we camped, hunting was not permitted within a half-mile radius, but in the mornings and evenings we often heard shots not far away. We decided that the safest time to let the Creeps fly was around the middle of the day, and as an extra precaution we allowed only three or four to fly at one time so that they were less likely to go far afield. If they began to fly too far we called to them at the top of our lungs.

'Creep! Creep, Creep! Come on Creeps!' we shouted, until they circled to land beside us.

Needless to say, any person passing by and hearing us shouting as we stared up into the sky, must have thought we were out of our minds. After the open expanse of the tundra, the Creeps now had to learn to avoid obstacles like trees, cabins and parked cars, especially when landing. One of the blue geese was our first casualty, damaging the leading edge of one wing on the sharp spike of a small broken-off tree. For a few weeks he couldn't fly, which earned him the name 'Slack Black', but in time he recovered completely.

Fred and Canada Goose were free most of the time, but once or twice when they flew off and did not return, we spent worrying hours searching for them. Once we discovered that Fred had decided to cut short his flight half a mile from camp to go berry-picking! He had developed a passion for the ripening blueberries and crowberries and was adept at picking them one at a time with the tip of his bill, then swallowing them with an almost imperceptible toss of his head. One scientist later remarked to us that Fred's bill represented the most efficient set of chopsticks he had ever seen.

We remained camped near Churchill for almost a month, chartering a helicopter when the weather was favourable and searching for the snow geese migrating along the shore of Hudson Bay. We also tried to approach the white beluga whales swimming in the bay and in the murkier water of the Churchill River, but they disliked the noisy machine and dived from sight before we could come close enough for filming. From the helicopter it was easy to spot polar bears, either along the rocky shore of Hudson Bay or farther inland among the trees growing around the shores of the numerous small lakes that dot the whole area.

Occasionally polar bears prowled around our camp at night searching for food, so we kept all our supplies locked inside the two vehicles, and the Creeps had to spend each night inside a nearby garage. We often saw polar bears along the stretch of coast between our camp and Chur-chill and each evening several visited the city garbage dump looking for scraps. Until ice forms again on Hudson Bay in late October or early November, the bears are unable to hunt seals, and live on a largely vegetarian diet.

From our camp near Churchill, we watched countless flights of geese pass overhead, and at night heard their calls as they flew towards the southern shores of Hudson Bay. Other geese, nesting farther north on Baffin Island and eastern Southampton Island, follow the eastern shore of Hudson Bay as they begin their southward migration, before breaking their journey to rest in James Bay. Some flocks, mainly of blue geese, then fly non-stop to the coastal marshes of Louisiana, covering the 1,500 miles in as little as forty-eight hours. However, most of the geese make a more leisurely journey, resting and feeding at stopover points along the migration route, in much the same way as an aircraft makes refuelling stops.

On their long migrations the geese fly along well-established routes known as flyways. Already we had become acquainted with the Pacific Flyway, to the west of the Rocky Mountains, but there are three other major flyways used by the snow geese in North America. The Atlantic Flyway is used by the greater snow geese which nest in north-western Greenland and in the high Arctic of eastern Canada, pause for six weeks each autumn on islands in the St Lawrence River, and then continue non-stop to their wintering grounds along the Atlantic coastal marshes from Chesapeake Bay to North Carolina. A few lesser snows fly with them, but the main populations of lesser snow and blue geese which nest in the Canadian north follow the Mississippi and Central Flyways, wintering along the coast of the Gulf of Mexico, in Louisiana and Texas, and some even crossing into Mexico.

8 Over the Border to Sand Lake Refuge

We planned to follow the snow geese using the Central Flyway, for although these birds come from several nesting colonies, they include the McConnell River geese. Our journey south had to be timed so that we would meet the migrating geese when they paused at National Wildlife Refuges in the Dakotas, Iowa, Missouri and finally Texas. Geese following the Mississippi Flyway do not stop along the way in autumn as much as the Central Flyway birds.

For us, travelling was now complicated by having to take our large family wherever we went, and our plans always revolved around what was best for the birds. From Churchill we travelled the first five hundred miles by train, with the Creeps riding in their travelling boxes in the baggage van. During the twenty-four hour journey to The Pas, we walked through the moving train several times to check on the birds and give them fresh water. The geese did not seem to mind the train trip, but Fred was decidedly upset, refusing all food and making plaintive little calls. By this time he had grown so much taller than the geese that we had bought a folding wire dog crate for him so that he could travel more comfortably. At The Pas we rested the birds for a few days before beginning the long road journey south in our two vehicles – the Chevy now pulled a trailer carrying the Argo and the Creeps.

One day Les was at a garage in The Pas waiting for a flat tyre to be fixed and fell into conversation with another motorist. Noticing Les's Australian accent, the man asked:

'What are you doing in this part of the world?'

'We're making a film about snow geese,' Les replied.

The motorist looked bewildered, then asked, 'What are snuggies?'

When Les told us the story that evening, we all had a good laugh, especially Lee, who blamed the misunderstanding on his Australian accent. However, a few weeks later, she was caught the very same way when a fellow American asked her what sort of birds we had in the boxes.

'Young snow geese,' Lee replied.

'Snuggies?' said the puzzled lady.

From that time onwards, we referred to snow geese as 'snuggies', and decided that blue geese must therefore be 'bluggies'.

We had difficulty finding somewhere safe to fly the Creeps at The Pas, for although there were no snow geese in the vicinity, the hunting season was in full swing for ducks, Canada geese and moose. Many people offered suggestions and we first let them fly on a farm some distance out of town, but there was very little cleared ground there for the Creeps to use as an airstrip. Some of them were so frightened when they landed among tall weeds that they just sat there partially entangled until we freed them and helped them out into the open. On the tundra they had never encountered anything like that!

It was a relief when Jack Lamb, President of Lambair, suggested we try nearby Grace Lake, which is a seaplane base and therefore closed to hunting. This proved to be ideal. Although the Creeps hated noisy aircraft, the lake was far from busy and we exercised the birds there several times. They loved flying out over the lake, passing in front of golden cottonwood trees as they banked to circle back to us. We also enjoyed Grace Lake, for there were many wild ducks on the lake and muskrat houses dotted the shallows. These rodents were particularly active during the autumn. Fred and Canada Goose flew with the geese, but we did not trust the highly-strung greater scaup enough to free them completely for exercise. The male had escaped from our lake-shore pen when we were camped near Churchill and we had some trouble recapturing him. After that incident, we trimmed a few flight feathers on one wing of each duck, so that they could still fly a little, but not too far. The scaup loved swimming and bathing. In fact, they frequently made us laugh by standing near the small dish of drinking water in their pen, and dipping and bobbing as they went through all the motions of bathing – on dry land! For safety at Grace Lake we placed their whole wire cage in the shallows so that they could enjoy them-selves in the water while the Creeps had their flying exercise. However, one day, as we lifted the cage from the water, the door catch slipped open, and away went our pair of scaup across the lake. Had we felt they had a chance of survival in the wild, we would have been more than happy for them to remain free. But with their partially trimmed wings they would be caught when everything froze in the grip of winter, and would face a slow death, if hunters didn't get them first.

After some coaxing, the male scaup came back towards us and swam about in the deep water ten yards from shore. Luckily, a muskrat unexpectedly approached the scaup, giving him such a fright that he swam back to us for help! Having caught the male, we borrowed a canoe from Lambair to paddle out on the lake, searching for the female. We had last seen her flying just above the surface with half a dozen other ducks. After several hours of fruitless searching, and with darkness descending, we had to give up for the day. Next morning we used the canoe again, although we were not very hopeful of finding her with so many wild ducks around the lake to confuse us. Just as we were about to give up the search, Jen thought she saw a scaup disappear through the entrance to a little bay which was almost hidden by the reeds, and sure enough, it was our missing female. Finding her was one thing, but recapturing her was quite another. It was fairly easy to corner her, but then whenever we paddled close she would dive right under the canoe. However, after half an hour of this game, Lee finally managed to catch her.

On a later visit to The Pas, outside the hunting season, we stayed with Bob Mushrush, at his lakeside Evergreen Lodge, the perfect place for the geese and Fred, as they could be free all day long. Bob had a large friendly dog, which Fred and the geese accepted after a very cautious first meeting. But there was another resident of Evergreen Lodge which had a very strongly developed hunting instinct, and delighted in stalking the geese. This was a tiny little black and white kitten, hardly big enough for a goose to stumble over. While the geese enjoyed themselves cropping the lush grass of Bob's lakeshore lawn, the little kitten would patiently stalk them from behind a bush. With tail swishing from side to side in anticipation, it finally made a charge into the midst of the unsuspecting geese, scattering them in all directions as they ran to take off, with the lion-hearted kitten racing along behind. This pantomime happened many times, but we laughed at the thought of what would happen if one of the geese had turned to face the tiny charging kitten and said the equivalent of 'boo'!

Travelling southwards from The Pas by road was a little like travelling with very young children, as our birds were entirely dependent upon us for all their needs. During the months ahead the Creeps were to travel over 10,000 miles with us. Although they proved to be placid, adaptable creatures, life on the road was not always easy, for them or for us. Overnight stops often had to be made at motels, and we always had to be careful to choose one which had a grassy area around the back, suitable for erecting a temporary pen where the birds could bathe, flap and feed. Before signing in at a motel, I would ask permission from the manager to have the geese at the back, and I'm happy to report that the answer was always yes. As a matter of fact, during our travels together the geese helped us to meet a lot of fine people. At night while we slept, the Creeps were back in the safety of their trailer home, for we were afraid that large dogs might break into their exercise pen. Canada Goose was something of a problem: less aggressive than the snows and blues and outnumbered by them, he was bullied if penned with the others in a small enclosure. On the other hand, if we put him in with the two greater scaup, he in turn attacked them. So spoilt Canada Goose had to have a pen and travelling box to himself.

Because our birds would need clearance from a U.S. Government Veterinary officer before being admitted to the United States, our permits stated that we must pass through the border post at Pembina, North Dakota. Arriving there one morning in early October with our passports and visas in order, we soon passed through immigration formalities. Customs officers cleared our vehicles and their contents without delay until they came to the birds. Officials carefully checked our sheaf of permits for the birds and found them in order, but it is not every day that hand-raised wild geese appear at the border for clearance. One Customs officer asked a lot of questions, thumbed through his books and phoned his superior, all the while trying to determine under which category our birds should be permitted to enter the United States.

'Mr Bartlett, could you place a value on these birds for me?' he asked.

'I'm afraid not,' I replied. 'We could never sell them.'

'But they must have some value,' the Customs man persisted. 'After all, you are using these geese in a film.'

'These birds really belong to your Uncle Sam, not to me,' I explained. 'When we finish making this film, we shall be handing the geese over to the U.S. Fish and Wildlife Service.'

Replying to questions from some of the onlookers, who were interested in the geese, we told them that the Creeps always came back to us after flying free.

'In fact,' I said, 'if we had released them in Canada just before driving across the border they would have flown over to join us in the United States. After all, the wild geese fly over and have no trouble crossing the border twice a year!'

This statement brought forth no comment, only disbelieving smiles from everyone within earshot. At this moment the Veterinary officer, Dr Allen, arrived on the scene to inspect the birds and was surprised to find how tame they were. He then conferred with the Customs officer and together they decided that the Creeps could be admitted to the United States under a section dealing with the importation of birds for scientific studies. Greatly relieved to have everything sorted out at last, I thanked the various officials and then asked:

'Would you like to see the geese fly free?'

'I would,' one Customs man replied.

While he held up the traffic, we released four of our geese. One of these happened to be Slack Black, who decided that his wing was not yet strong enough to fly, but the other three took off into wind and made a beautiful circuit of the border post. They were all set to carry on flying, but when we called to them they obediently landed alongside us on the broad roadway. We were very proud of our charges for performing so beautifully and everyone seemed most impressed with the flying display. In fact, one Customs officer even asked for our autographs!

Once through the border we set off on the trail of the migrating wild geese. We paused briefly in North Dakota to drop off the pair of greater scaup at the Northern Prairie Wildlife Research Center, near Jamestown. If the Creeps chose to remain with us until we completed the snow goose film and left the United States, this would be their future home also.

Our first major stopping point on the Central Flyway was Sand Lake National Wildlife Refuge, thirty miles from Aberdeen, in north-east South Dakota. It had taken us four days to drive the 900 miles from The Pas, and during the last two hundred miles we frequently saw long V-shaped formations of snow and blue geese heading southwards high overhead. We had timed our arrival at Sand Lake well. It was 13 October, and there were an estimated 60,000 snow geese on the refuge, with more arriving every day. We saw comparatively few Canada geese but there were large numbers of mallards and pintails, together with many less numerous species of waterfowl.

The refuge manager at Sand Lake, Lyle Schoonover, had been there for over ten years and was dedicated to geese. Not content with a normal eight-hour working day at Refuge Headquarters, Lyle 'lived geese'; they were his life. So he was not at all put out when we arrived on his doorstep with our family of tame geese and Fred.

Following Lyle, we were soon shown a large roofed-over enclosure, complete with a big wooden shelter and running water in the pool.

'We used this for our captive flock of Giant Canadas during the winter,' Lyle explained. 'The water comes from an underground bore and doesn't freeze. I'll get some of the men to clean it out and you can use it for your geese: we don't need it ourselves until the end of November and I imagine you'll be gone before then if you're following the wild geese.'

The Creeps immediately had a thorough bath after the long journey. Fred was in his element, probing the ground with his long bill or dancing up into the air with broad wings spread. In such a large pen it was safe for Canada Goose to be with the others, as they had plenty to keep them amused without pecking at him. Until now the Creeps' diet had consisted of the Miracle dog food and greenery, but at Sand Lake it was not long before Lyle Schoonover introduced them to fresh young corn on the cob. For the Creeps it was love at first taste and even Fred enjoyed the tender young kernels – but only after they had been cut off the cob for him. The geese soon developed a liking for dry corn, wheat and barley, but Fred would never touch such dry 'inedible' stuff, though wild some cranes feed on grain for part of the year.

With the Creeps settled, Lyle took us for a tour of the area, filling us in on the history of Sand Lake Refuge and its present status. It was not until 1935 that the wildlife refuge came into being. Two low dams were built and now half of the refuge's 21,451 acres consists of excellent water and marsh habitat that is enjoyed by a variety of migratory birds. The remaining acreage contains fields of corn, barley and rye, grown especially to attract migrating waterfowl in autumn, as this helps to keep the birds away from the surrounding prairie farmlands until crops can be harvested. Actually it was not until the 1940's that snow geese began using Sand Lake Refuge as a resting point on their migrations. At first only a few hundred geese were seen here each year, but this has gradually increased until a peak autumn population of as many as two hundred thousand geese might be counted after a good breeding year in the Arctic, with more than half the total being made up of young birds travelling south with their parents.

There are two things the geese need to attract them to an area. The first of course is food, but there must also be a fairly large area of water nearby where they can bathe and rest unmolested. At Sand Lake Refuge there are almost non-stop flights of geese on the move all day long as they shuttle back and forth between the lake and the fields of grain. Thousands of geese feed together, taking only a day or two to clean up the grain in each field before moving on to another part of the refuge, or flying out to surrounding farms where the harvesters have left behind waste grain.

Soon after reaching Sand Lake we watched snow and blue geese massed in a large field on the refuge, where a corn picker moved noisily back and forth harvesting the corn. The geese had become so accustomed to the harvesting machine that they would often allow it to pass within fifteen yards of them. But if a person on foot tries to approach the feeding geese they all take off before he comes within a hundred yards. From a safe distance we watched and filmed the activity in this field for an hour or two. Then as the picker passed them some of the geese took fright and flew into the air calling loudly, alarming the entire flock. In seconds the air became so filled with flying geese that the machine working behind them was completely blotted from our view for perhaps half a minute. It was as if a thick carpet of geese had lifted off the ground, not all at the same instant, but forming a continuous wall of birds rising almost vertically; those nearest the machine took off first, followed by their neighbours until they were all high in the air and we could once again see the harvesting machine moving steadily on its course. A few of the flying geese headed for the lake, but most circled back past a backdrop of yellowing cottonwood trees to land in another part of the field to carry on feeding. Scanning the immense flock with binoculars, we were pleased to see a few of the geese wearing the orange neckbands we had seen fitted at the McConnell River. It was like meeting old friends again.

In spite of the worsening weather, we decided that we would camp in the refuge. We would then be close to the Creeps and also be able to make the most of all filming opportunities. We were not exactly looking forward to the prospect of camping in tents again in the cold weather, but Lyle Schoonover came to our rescue.

'I've had an idea I think you might like,' he said as we stood together watching the Creeps.

We followed Lyle to a small building which he referred to as the Bird House, as scientists studying birds often stay there during the summer months. Lyle assured us that we were welcome to use this little converted barn for as long as we were filming at the refuge, but apologized for its Spartan appearance. Equipped with bunks, a bathroom, stove, heater and refrigerator, the Bird House looked like a palace to us after five months of camping and a week or two in motels. If the weather was bad we could work indoors, yet we were always on hand if something of interest happened. Lyle could not have been more helpful. He even insisted that we help empty his well-stocked vegetable garden before the ground froze, and after our months in the north we really appreciated the fresh garden produce. We had become so involved with the geese that we referred to ourselves as 'Goose Nuts', and soon found that Lyle had been one for many years. We spent countless absorbing hours with him, often until after midnight, talking about geese and learning a great deal.

Among other things, Lyle told us that for the past twenty years at Sand Lake Refuge, about one thousand snow and blue geese have been caught each autumn for banding. At a permanent site near the lakeshore, corn is used to bait the geese so that they feed at the same place every day. The grain is spread on the ground alongside carefully folded nets, which are attached to eight-pound projectiles, four to each net. When detonated, an explosive charge propels the heavy projectiles, pulling the nets high above the feeding geese.

On a calm morning in late October, we waited with Lyle Schoonover and five other refuge men, in a sheltering belt of trees two hundred yards from the cannon-net site. The dynamite plunger was connected to wires running out to the cannons, then Don Snider, one of the refuge staff, climbed a tower amongst the trees to watch the geese through binoculars. When enough were massed in front of the nets he gave Lyle the signal. As the plunger was pushed home, the nets streaked out over the unsuspecting geese before the bang of the explosives reached our ears. With a roar of wings hundreds of geese lifted into the air, calling excitedly as they flew out to settle on a nearby lake, leaving behind a few hundred birds trapped harmlessly beneath the nets. The refuge personnel carefully removed the geese, sexing and banding each one before turning it loose to fly off unharmed to rejoin the rest of the flock.

Three firings of the cannon-net may be sufficient to catch the total of one thousand geese required for banding each year, but this work must be completed before the weather becomes cold enough to freeze the nets to the ground. Earlier each season, before the geese arrive at the refuge, one thousand mallards are caught for banding at the same cannon-net site. Each year the men also try to band five hundred Canada geese, but had no luck the year we were there as the Canadas remained scattered. We watched the cannon-nets operating at Sand Lake Refuge many times, and not a single bird was injured.

For the Creeps, life at Sand Lake was a far cry from the complete freedom they had enjoyed around our tundra camp, but it was certainly the next best thing. Each day we let them out, a few at a time, to fly free. Unfortunately we could not explain to the Creeps that they were likely to be shot if they flew beyond the road that marked the limits to the refuge, less than half a mile away behind a belt of trees. Fred often flew with the geese, but was always satisfied with one or two wide circuits, landing beside us long before the Creeps had finished flying. When the three or four geese landed, we walked them over to a second pen and closed the gate behind them. Then we let a new group out to fly, moving them to the other enclosure when they had had enough flying exercise. By doing this, we made sure that every one of the Creeps flew each day, and the two adjacent pens were ideal for this purpose.

When flying free at Sand Lake the Creeps often passed over a large pond below their pen, where wild geese mingled with the captive flock of giant Canada geese that were kept for breeding purposes. Occasionally some of the Creeps would land to visit their wild cousins. We then spent a worrying half-hour or more, until finally they responded to our calls and came back. If Fred happened to be with the Creeps when they flew near the wild geese, the latter panicked and took off in alarm. Cranes normally fly with their long legs trailing behind them, but on one particularly cold day we were amused to see Fred flying with only one leg trailing and the other tucked up against his body for warmth! He certainly felt the severe cold more than the geese and often stood on one leg, with the other completely hidden among the feathers on his belly. On cold mornings, we frequently had to break the ice on the surface of the Creeps' pool and then they would drink and swim happily in the freezing water. It was painful for us even to dip our fingers in such chilling water, but the geese loved it. Apart from a goose's bill, the only parts unprotected from the cold by layers of down

Jen and Des in camp with the imprinted geese
and the two sandhill cranes, Fred and Lady

The imprinted goslings follow Jen and Des across the tundra
The lion-hearted kitten at Evergreen Lodge puts them to flight

At Sand Lake Refuge they would fly for over a mile behind our car

Fred giving a warning call

Polar bear visiting our camp

The immaculate black and white plumage of an Arctic loon with its chic

A confrontation between a pair of nesting sandhill cranes and
a family of snow geese

and feathers are the legs and feet, and these have a complex network of tiny blood vessels, which enable the birds to stand on ice and to swim in very cold water without losing much body heat.

As snow geese are birds of the wide open spaces and like to have an unobstructed view of their surroundings, it was thought that they would never feed in standing corn. However, hunger must lessen caution, for several years ago the geese suddenly began feeding on the standing corn within the protected boundaries of Sand Lake Refuge, and have continued to do so each autumn. It is fascinating to watch them fly above one of these fields, reducing forward speed to hover just above the standing corn, then expertly helicoptering down to land in the narrow corridors between the rows of tall plants. All summer the geese have been limited to a diet of tundra vegetation and now they are determined to get at the corn, even though the ripened cobs are attached to the plants as high as three and four feet from the ground and still encased in their stiff husks. By jumping up, the geese are able to pull some of the cobs to the ground, and strip others of the ripened kernels as they hang on the standing stalks. Once they find themselves a safe area the gregarious snow geese will continue to return to the same place day after day, until all the food has been consumed.

With the idea of setting up remote-controlled cameras along the outermost row of corn, we once approached a large field of standing maize where the geese were massed feeding. The geese close to us took off and flew to the far side. As we set up the complicated equipment, however, we soon noticed that the sounds of feeding geese were coming closer and closer as others worked their way towards us, oblivious of our presence. Although we couldn't see them through the dense wall of tall corn plants, some of the geese must have been only fifteen yards away, judging by the crashing sounds they made while pulling at the dry husks. But the most surprising thing was to find that at close quarters the murmur made by the thousands of feeding geese resembled the humming of a gigantic beehive. From farther away, this murmuring sound is completely drowned by the more strident calls of other geese overhead, as they fly noisily in and out of the field.

We were amazed to see just how quickly and thoroughly the flocks of geese flatten a crop of barley once they move into a field and begin feeding. On the north-west corner of the refuge we watched them reduce one huge barley field to mashed straw on bare ground in two days, for at times there must have been 40,000 geese feeding within a quarter of a mile of where we sat in the Land-Rover. Late one afternoon the massed geese were still feeding in this field as the sun went down, a molten red ball sinking behind a row of now leafless trees. But on this particular evening we were not trying to film sunsets. For several days we had been watching the moon and knew that on this night it would rise soon after the sun had set, before complete darkness had fallen. The sky was clear and only ten minutes after the sun had set, we were rewarded by the glorious sight of a huge orange full moon lifting slowly above the eastern horizon. Small formations of geese passed in front of the moon as they flew in from the lake to join the others already feeding on the barley. Normally most geese stop feeding and return to the safety of the lake by nightfall, but these seemed to be taking advantage of the bright light from the full moon. Suddenly, something unseen by us frightened the geese. With a tremendous roar of wings followed by a deafening clamour of many thousands of excited voices, they rose into the air. Silhouetted against the pale mauve of the evening sky their numbers almost completely eclipsed the moon before they circled to land in the field again. Chilled to the bone, we finally left this enchanting scene, while most of the geese were still feeding in the bright moonlight.

During the autumn, Sand Lake Refuge is closed to visitors, except for the main county roads which cross part of the reserve. The reason for the visitor ban at this time of year is because of the hunting season. Sand Lake becomes a true refuge, for there is no one to scare the geese into flying out past shotgun-toting hunters waiting beyond the boundary.

However, after spending the night safely on the surface of Sand Lake, many of the geese do fly out of the refuge to feed on waste grain in harvested corn fields on farms up to twenty miles away. Somehow most of the geese seem to know, perhaps from experience, that they must fly high to avoid the hunters whenever they leave the refuge, and their formations usually gain a few hundred feet of altitude before crossing the boundary. Remaining at this safe height until they are directly over their objective, they spiral down to land and feed. The fact that they feed gregariously helps during the hunting season. Many thousands of eyes watch for danger, and should a hunter or two manage to crawl within shooting range of a flock of several thousand geese only a very small proportion of the birds will be killed before the others fly out of range. If only one family of geese had been feeding in this field, it is very likely that they would all be killed.

A true hunter goes to a lot of trouble constructing a well-concealed blind. On the flat prairies of the Dakotas this is often a deep pit dug in the middle of a harvested corn field, or perhaps two or three pits twenty yards apart, so that a group of hunters can hunt the same field shortly after dawn. The soil from the pits is carefully driven away, and corn stubble scattered on the wire screen roof camouflaging the centre sections of each pit. Two hunters can then operate together, one at either end of the pit: if there are three pits, six hunters can shoot at the same time. Before hunting can begin, the geese must start feeding in this particular field, which in South Dakota would be one section – one mile square, or 640 acres. The owner of this field then has to police it, keeping other hunters out so that the geese are not frightened away on the first day they begin feeding there.

That night the telephone at the owner's house will be busy, as hunting friends are invited to participate in a snow goose hunt the following day. In the cold pre-dawn darkness the men meet around 4 a.m. to walk out to the pits. Goose decoys are set up around the blinds, about a hundred being a suitable number. But the hunters can increase them by the simple technique of throwing large bundles of white paper plates high in the air, and allowing the wind to scatter them in all directions amongst the decoys. As the eastern sky turns pink, the hunters take up their concealed positions in the pits, having first received careful instructions from the leader.

'Remember, I'll be using the goose caller and no one is to shoot until I give the word. When they are low enough and close to the pits I'll shout "Take them!" Stand up quickly then, and good shooting!'

The trained bird dog also jumps down into the pit, following its owner. There is a whistle of wings as a flock of ducks passes over, black silhouettes against the sky. But the men are waiting for geese. By law, hunting cannot start until half an hour before sunrise, ceasing again at sunset, and the state game department publishes the exact times for hunting on each day throughout the season. No hunter can therefore claim that he did not know the time to begin shooting if he is apprehended for hunting too early in the day.

In the distance, flying snow geese can be heard calling to each other within the flock, and the pulse of the waiting men quickens. Skilfully using the caller, the hunter mimics the sounds of feeding geese. Soon the birds are within sight of the crouching men. Flying high, they begin to circle as they spot the flock of decoys down below. Lower and lower they come. The flocks are soon concentrated above the corn field as more fly in higher up, and the spiralling geese descend. With wings set, they have decided that it is safe to land among the 'geese' already feeding on the ground.

'TAKE THEM!'

Six hunters stand with shotguns blazing. Flying geese execute lightning-swift flight manoeuvres, as powerful wings beat frantically to climb out of range. For the lower birds the end comes swiftly and without warning, several falling to the ground during the first fusillade.

At a shout of 'Fetch!' the retriever is out of the pit in one bound followed by the hunters. All the dead birds are collected, and cripples quickly dispatched. Fifteen snow geese all told; half the bag limit allowed the six hunters for one day; a particularly productive shoot. The dead geese are placed out among the decoys and the men vanish once more into the concealed pits. Before long, if they are lucky, there will be more geese passing over. After six more shouts of 'TAKE THEM!' the hunters call it a day – their total kill, twenty-seven geese. Another shoot, and they could easily exceed their daily limit.

Of the twenty-seven snow and blue geese shot by this group of hunters, twenty-one are juveniles, very nearly eighty per cent of the total. This is normal early in the season, as the young birds have not had time to learn the dangerous ways of man. Also, with the disruption of family groups during the hunting season, the juveniles decoy readily as they search for missing parents.

The type of goose hunting just described is the good clean hunting favoured by the game departments in the various states. If they are lucky, these hunters would have one, or perhaps two more hunts in that particular field. The geese are given complete protection during the day following a dawn hunt, and in a few days all the waste grain is eaten out and the same geese may be feeding in another field perhaps ten miles away.

On the perimeter of Sand Lake Refuge there has been another type of hunter, perhaps more numerous, who shoots at the geese as they leave or return to the refuge. On an earlier visit to Sand Lake in October, 1969, we had been amazed to see hunters lined up in the ditches beside the main road running along the refuge boundary. These hunters often shredded the overhead electricity wires as they fired wildly at passing geese, and the practice seemed to be to use more powerful ammunition with buckshot for the high-flying birds. So the crippling rate was very high, many wounded birds continuing to fly on with the flock. Any goose that fell inside the refuge boundary after being shot could not be retrieved, since the hunters were not permitted to climb over the fence. On our more recent visits to Sand Lake we were relieved to find that hunting from the roadway is prohibited. Hunters who do not have access to private farmlands can use the two hundred, or so, blinds spaced along certain sections of the refuge boundary, with a retrieval zone in front of the blinds. These blinds are occupied on a first-come basis, so most of the hunters sleep alongside them overnight, especially on week-ends.

Many of the injured geese fly back into the refuge where the refuge personnel pick up as many of the cripples as they can and place them in a large open enclosure with plenty of food and water. Those that completely recover are free to fly out and rejoin the wild flocks, while others that will never fly again are given good homes in parks and reserves. Any grounded cripples that remain free face a slow but sure death after the rest of the wild geese move south, as the lake freezes over and any available food is gradually covered by snow. The family ties of the snow geese are so strong that the rest of the family will remain with a crippled member long after they would normally have flown south, only leaving the injured one behind when their own survival is threatened.

While we filmed at Sand Lake in the autumn, crippled geese were a daily sight and it is impossible not to be saddened by them. For a day in late October our diary records a typical incident: 'Saw several cripples today. One young snow with a newly bloodied left wing sat on the road with two adult snows. As our Land-Rover approached slowly the two adults took off but when the wounded one tried to follow, it performed a cartwheel and landed on its back, rolling into the grassy ditch beside the road before walking off towards the lake with the broken wing dragging awkwardly. We were able to catch this cripple and took it back to the pen at refuge headquarters.'

The autumn hunting limits are set after the nesting productivity of the snow geese for that particular year has been estimated. The daily limit is normally five geese per hunter, but was reduced to four following a poor year on the major nesting colonies. Usually there is a limit of one Canada goose per hunter. When we were in South Dakota the goose hunting season extended from 1 October until 11 December, which more than covered the time that the geese spend in the state. In Texas, where geese arrive much later, the season ran from 3 November to 16 January. Every hunter has to buy a state waterfowl hunting licence as well as a Federal Migratory Bird Hunting Stamp, commonly called a Duck Stamp. There are two million of these Duck Stamps issued annually at five dollars each and the proceeds go into a fund set aside for the acquisition of wetlands for migratory birds. Needless to say, this is a major contribution towards ensuring the future of the geese and other migratory birds, and even though the hunter's main interest may be to ensure that there will always be birds enough for him and his descendants to shoot at, we doubt whether any other group of two million people would voluntarily give five dollars apiece each year purely towards conservation of waterfowl.

Although it is difficult to arrive at an accurate figure, close on half a million snow and blue geese are killed by hunters each year on the North American continent. Provided the geese have successful breeding seasons on the tundra, biologists feel they can maintain their numbers even if twenty per cent of their total fall to hunters each autumn.

9 Moving down the Central Flyway

Towards the end of October the first snowstorm hit Sand Lake Refuge, although there had been a few dustings of snow earlier. It began with freezing rain and sleet, followed by a few inches of snow. The day following the storm dawned completely clear with the temperature not much above zero degrees Fahrenheit. We stepped outdoors that morning into a fairyland of white where every twig and berry, every blade of grass and the wire of every fence was encased in a thick layer of clear ice that sparkled like thousands of diamonds in the bright sunshine. This was our first experience of an ice storm and we could not imagine anything more beautiful, although we learned that the sheer weight of the ice frequently causes damage to trees and power lines. In spite of the cold we remained outdoors most of the day enjoying the spectacular scenery and watching the geese search for waste grain in the snowy fields of corn stalks. With the very cold weather much of the lake had frozen over, leaving only small patches of open water where the waterfowl could drink and bathe. Many snow and blue geese, together with some Canada geese and mallards, stayed far out on the ice near open water, while others rested in a bare snowy field that had once contained barley. Many of the geese had moved south ahead of this storm and others departed soon after the weather cleared. While there is still open water and the snow is not too deep in the fields, some of the geese are content to linger, but by mid-November virtually all had left South Dakota and it was time for us to follow them south.

Once again we loaded our equipment and the Creeps into the two vehicles and drove southwards. High overhead we saw the long diagonal lines of V-shaped flocks travelling in the same direction, and at night we could hear them calling to each other as they flew onwards. An average flock contained between two and three hundred birds and before one formation was out of sight new ones were passing overhead. Most of the geese flew a thousand feet or more above the ground, but from aircraft in Canada they have been observed flying as high as six thousand feet or more. Usually they wait for favourable weather conditions before commencing a long flight and can vary their altitude to take advantage of helpful tail winds. These long flights often take place at night, and although many brilliant scientists have studied the mysteries of bird migration there is still much to learn before we fully understand how they manage to navigate so accurately. The stars, the sun and the earth's magnetic field all play their part, but it seems likely that for years to come, a certain amount of mystery and romance will continue to surround the incredible flights made by snow geese and many other species of birds.

After two days of driving we reached De Soto National Wildlife Refuge on 16 November. Located on the Missouri River where Iowa and Nebraska meet, De Soto is little more than a third the size of Sand Lake Refuge, but we arrived there to find that an estimated 300,000 geese, mostly snows and blues, were using the refuge. This area was only declared a refuge in 1958 and before that time very few geese actually stopped here. Some crops are grown on the refuge, but the majority of the geese make morning and evening flights to the surrounding farmlands, often many miles away. Returning from these feeding forays they are attracted to a wide oxbow of calm water that has been cut off from the main flow of the Missouri River. We had seen massed snow geese at Sand Lake, but here at De Soto the flocks were bigger and the birds seemed to be even more densely crowded together on the oxbow and along its banks.

We noticed that the geese had a favourite resting area on the bank and this gave us an ideal opportunity to set up our radio-controlled cameras under snow goose decoys which Les had made up from geese found dead on Sand Lake Refuge. As we advanced slowly to arrange our equipment the geese retreated into the water and moved a couple of hundred yards away. We set up one movie camera and one still camera at ground level, each camouflaged by a snow goose decoy with the lens pointing through a hole in the breast. After covering the radio equipment with a third decoy, the connecting wires between the three

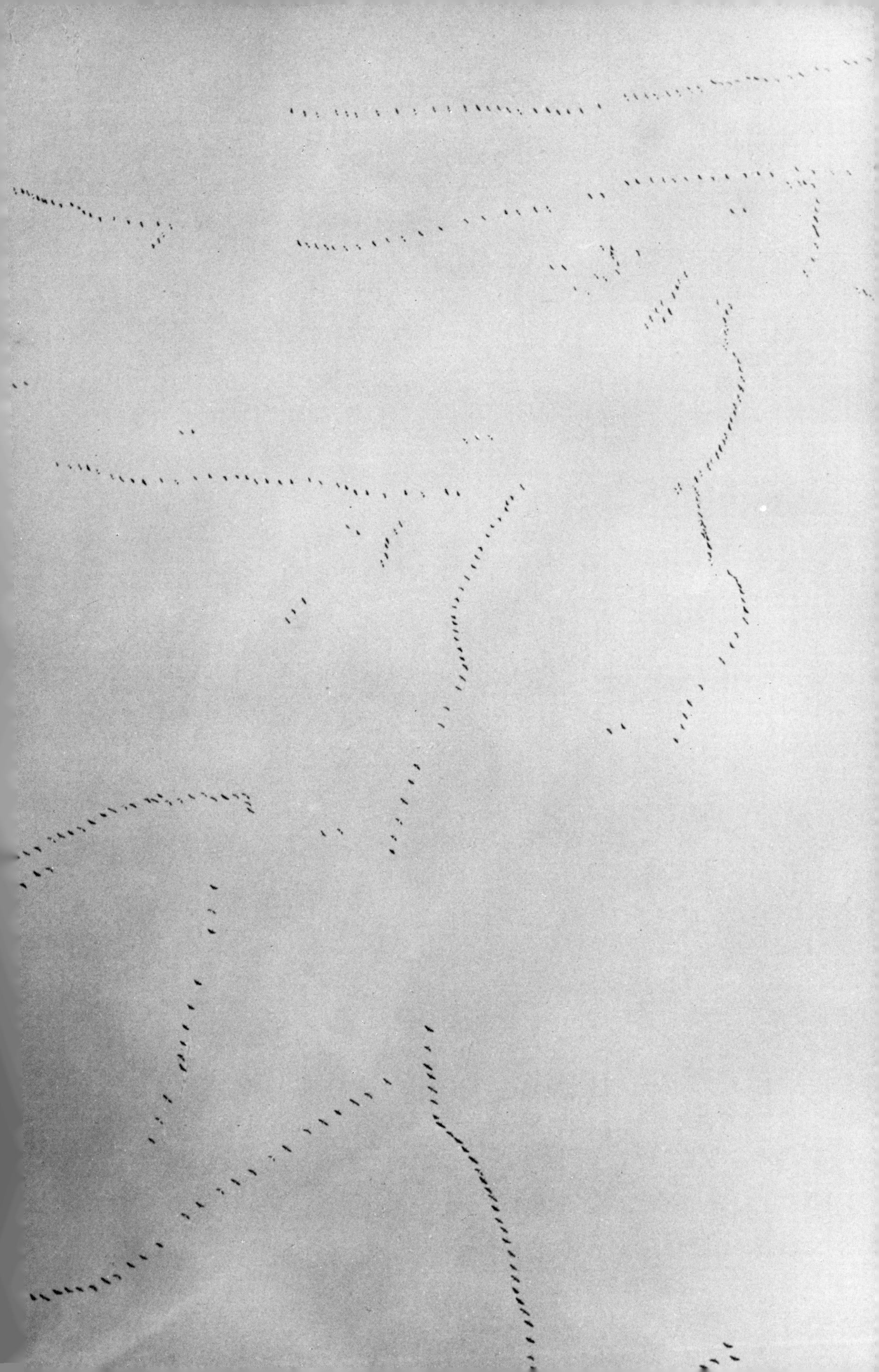

Snow geese form ever-changing patterns in the sky

The 'Creeps' travelling in their trailer

U.S. Customs, Immigration and Veterinary officers watch them before their flight into Canada

Snow geese landing in the Sacramento National Wildlife Refuge, California

dummies had to be buried so that the returning geese would not chew on them. Moving back three hundred yards we settled down to watch from the Land-Rover.

It was a brilliantly sunny day but the air was extremely cold and our fingers were numbed after setting up the intricate equipment. At first, we didn't think our goose-like cameras were going to be a great success. Patiently we watched and waited, but after an hour no geese had ventured within a hundred yards of the decoys. Then we had unexpected help from a high-flying jet aircraft which disturbed a huge flock of geese that had been resting farther along the oxbow, out of our sight. We heard their massed take-off and seconds later flying geese appeared above the trees, many of them heading in our direction. Within a few minutes they began landing on the surface of the water near us and along the bank of the oxbow. A real blizzard of snow geese descended until, even through binoculars, we could not identify our decoys among so many wild birds. Although we knew our cameras were surrounded by geese, working them by remote control was very much a hit or miss affair. When the film was processed we half expected to find part of a goose blocking off the scene as it slept through each and every picture we took. But it must have been our lucky day, for the colour photographs show an unusual wide-angled goose's-eye view of the scene within the flock, as some birds rest and preen, while others land nearby.

We had just calculated that we must be out of film in both cameras, when there was a tremendous commotion as perhaps 60,000 geese took off at once, leaving our three lonely decoys on the bank. At first we could not see or hear anything that might have disturbed the geese. Peering skywards through binoculars Jen at last solved the mystery.

'Look up there,' she exclaimed. 'It's that lone bald eagle that scared them!'

Soaring effortlessly five hundred feet above the oxbow, the circling eagle slowly moved away and most of the geese turned back to land on the water. The flocks of geese were equally concerned whenever an aircraft appeared anywhere in the sky, obviously mistaking it for another bird of prey.

From the Creeps we learnt that snow geese have much keener vision than humans. Tilting its head on one side to look skywards, a goose can spot a high-flying bird that is invisible to our eyes without the aid of binoculars, and while doing this the goose is not at all worried if it gazes directly towards the sun. Usually ceaseless chatterers, the Creeps would occasionally fall suddenly silent. We soon came to know that this meant they were intently watching a potential enemy, usually a bird of prey. If the hawk or eagle swooped too near, the Creeps always moved close to us for protection, even when they were full-grown.

At De Soto we kept the Creeps in an empty grain silo at the edge of a large field, letting them out to fly once or twice a day as usual. For the first few days this worked well; the Creeps were quite satisfied with a few circuits of the field. Then came a bitterly cold, overcast day. Because of the strong winds we cautiously allowed only two or three of the Creeps to fly at a time that morning. But when we released them for a second flight in the early afternoon, five snows managed to slip out of the pen together and immediately took to the air with Fred. Circling the field four times while gaining altitude, the geese then flew out of sight over the trees, while Fred turned back to land beside the silo. The five Creeps were obviously excited by the windy conditions and by the many flocks of wild geese flying high above. A few moments later they flew into sight again, calling excitedly to their friends in the enclosure below, before disappearing once more beyond the trees. A long five minutes dragged by before they appeared overhead once again, just as a noisy aircraft approached, raising many thousands of wild geese from other parts of the refuge as well as frightening the five airborne Creeps. They vanished from sight among the mass of flying snow geese and did not return.

Until darkness fell we took it in turns to stay by the silo in the freezing wind in case the truants reappeared, for if they did come back we wanted to be sure there was a welcoming committee of at least one to greet them and make sure they did not fly off again. Although it seemed hopeless to search for them among the vast flocks of wild geese, we could not bear to sit still and do nothing. Perhaps they had landed somewhere on the refuge with the wild geese, but how does one go about locating five unmarked young snow geese among flocks totalling a quarter of a million or more? We could only hope that they might recognize the Land-Rover or hear our calls. Les took first turn at the silo while Jen, Lee and I drove slowly around the refuge roads pausing near each flock of geese. After looking closely at them through bino-culars we called at the top of our voices,

'Creep, Creep. Creep, Creep. Come on Creeps!'

The wild geese all put their heads up alertly wondering what the fuss was about and many called excitedly, but none flew over to join us. As we sat watching one flock of geese on the oxbow we noticed several white-fronted geese among the snows and blues on the nearest bank. Known as a speckle-belly, because of the irregular black markings on its underparts, the white-front also breeds on the Arctic tundra though we had not seen any of their nests. In fact this was the closest view we had had of these geese and as we eagerly studied them through binoculars we noticed, just beyond them, a snow goose wearing one of the orange neckbands from the McConnell River. While it was good to know that this was one goose we had seen before, we would much rather have spotted our five missing Creeps.

As we drove slowly from one flock to another, a white-tailed deer occasionally trotted off unhurriedly from beside the road with its huge white flag of a tail in the air, switching from side to side. Here and there among the leafless tops of the sixty-foot-high cottonwood trees sat an eagle, ever on the lookout for a dead or crippled goose to dine on. We searched and called until we were hoarse and half-frozen, but finally had to give up when the light began to fail. Before leaving to spend the night at a nearby motel we paused at the refuge office to tell our sad story, just in case any of the men happened to see five unusually tame geese while going about their work.

The following day we were busy filming most of the time, while Les and Lee took turns staying with the Creeps at the silo. We had almost given up hope of ever seeing the five missing Creeps again when one of the refuge men approached the silo late in the afternoon carrying a wounded young snow goose in his arms.

'Do you think this could be one of yours?' he asked, and went on to explain how he had found it.

While grading a back road on the refuge he had noticed three geese huddled under some bushes and gone over to check on them. Two of the geese moved off as he approached but the third one seemed amazingly friendly and he felt it must be one of ours. And indeed it was! Apart from a bloody patch of feathers on his right breast he seemed to be fine. While we examined the wound made by the shotgun pellet he was completely docile but as he did not seem to be seriously injured we decided to let Nature take care of the healing. After examining this bird closely we were convinced that it was 'Clark Gable'. The only one of the Creeps that we had named while still on the tundra, he always seemed to be right in front of the camera when they were being filmed.

At this moment Clark was a very subdued goose but there was no doubt that he was pleased to be back. He soon recovered completely and in the months ahead became the undisputed leader of our little flock. Because of his adventure we tended to favour Clark and this no doubt helped him to attain his position at the top of the pecking order. But with his personality we felt he would have held this position in any case.

Of the other four truants we saw nothing more although we remained at De Soto for several days to see if they would reappear. Growing up with us as their foster parents, the Creeps had of course no reason to think of humans as anything but their friends. Now we could only hope that the missing ones had stayed with the wild flock, learning from them to fear and avoid man. We suffered from considerable feelings of guilt at having made the Creeps so trustful of human beings and it did little to ease our minds when we tried telling ourselves that most of our goose family would not have survived at all if we hadn't adopted them in the first place. We knew that we had been taking a calculated risk in letting the Creeps fly free, but this incident made us decide that the ones that were left would have to remain grounded until we could move away from the goose-hunting areas along the Central Flyway.

Saddened by this affair we went on south again following the wild geese to Squaw Creek National Wildlife Refuge in Missouri. It was a half-day drive by road for us from De Soto but a relatively short two-hour flight for the geese. Already 200,000 geese were using the Squaw Creek Refuge, with more arriving daily from the north, although the influx of new birds was more than balanced by the departure of other flights continuing their migration south to Texas.

It was 22 November when we arrived at Squaw Creek. As the geese had already eaten most of the grain on the refuge, they fed in surrounding farmlands, but returned in the late morning. Flights coming in from the east stayed high to clear the two-hundred-foot loess hills bordering the refuge and then spiralled down for hundreds of feet to alight on the water below. Many of the geese tumbled spectacularly downwards, losing altitude very quickly by sideslipping. It is as if the bird has completely lost control and is falling through the air like a large leaf. Texans have invented a special word for this manœuvre – they describe it as 'whiffling'. The action takes place far too quickly for the eye to follow in detail, but slow-motion film with two hundred pictures taken every second shows that the geese not only turn on their sides with wings fully open when whiffling, but occasionally roll completely on to their backs before righting themselves. They never go through a complete 360° roll, but the most incredible thing we learnt is that although their bodies may turn through 180° to achieve the upside-down flying position, their heads always remain the right way up in relation to the horizon. When standing on the ground a goose has no trouble turning its head 180° to look directly behind, but to us it seemed amazing that in mid-air a goose could turn its entire body and wings a corresponding amount while the head remained in a fixed position. Geese normally fly with their legs and feet held straight back, pressed closely against the underside of the tail. But our slow motion film shows that immediately before commencing to whiffle a goose suddenly brings both feet forward, bending the legs and holding the curled up webbed feet close to the body. This action no doubt alters the bird's centre of gravity and gives it greater control for aerobatics.

At first we thought that whiffling was something practised only by experienced fliers, but on examining photographs taken on the tundra we found that at least one of our young geese had been caught in the act of performing this manœuvre, only a few feet above the ground, during the very week when they first began flying! To us, this is astonishing. It is definitely instinctive and not something learnt from their parents. Observing the Creeps, we found that they whiffled more frequently in windy conditions and often performed as they flew over us. They usually called excitedly at the same time, and we could never be sure if the whole performance was just a sign of high spirits or if it was their way of showing off their flying ability to us. Or could it be that they were wanting to encourage us, their parents, to join them in the air?

During our week at Squaw Creek Refuge we had very little sunshine, with grey skies, fog or cold snowy weather much of the time. When some 50,000 geese were huddled at rest on the water during one snowfall we found they allowed us to approach quite close in the Land-Rover. This was definitely not flying weather, even for snow geese. After watching this huge flock at rest for some time we were able to single out three of our old friends, the geese marked with orange neck bands from the McConnell. We also identified several of the smaller Ross's geese, though it is only at close range that one can distinguish the characteristic warty protuberances near the purplish base of its small bill.

'We control the level of the pools to attract early loafing geese,'

Harold Burgess, the Refuge Manager, told us, 'but the mud banks and shallow waters freeze first as winter approaches. This encourages the waterfowl to continue their migration south.'

We were lucky enough to find accommodation in an old schoolhouse that had been converted into the Sportsman's Lodge, and the Creeps were given the use of a long empty grain shed at refuge headquarters where they were safe at night. As usual we made a wire pen outside on the grass for their daytime use. After a few days at Squaw Creek we noticed that for some reason Canada Goose had become very nervous. Then one morning when we arrived to let the geese out of the shed we found him in a terrible state. He was standing dejectedly, his feathers were very bedraggled and bloody patches of skin showed on his body. We realized that for several nights the other geese, and possibly even his friend Fred, must have been harassing the more timid Canada Goose. In the long, narrow shed they had kept him away from the end where we left food and water and had given him no rest until finally they wore him down and pecked him unmercifully. Luckily we rescued Canada Goose just in time and although he was in such a sad-looking state he responded rapidly to our care and attention, and recovered before the day was out.

Harold Burgess and his wife Ruth were kind enough to invite us to share their traditional Thanksgiving dinner which was the first family style, home-cooked meal we had eaten for many months. Fred the crane also scored at Thanksgiving. He had developed a liking for cooked chicken, which became one of his favourite foods, and he thought the turkey scraps were delicious!

On a dull day in late November we drove thirty miles north with Harold Burgess to check whether many geese were using the two small lakes on Opp's Farm, a private refuge where the birds are safe from hunters. There were relatively few geese on the water, but in the late afternoon high-flying formations of snow and blue geese began passing over, heading southwards, long lines and Vs of geese stretching across the grey sky from one horizon to the other. So closely did the flocks follow each other that it was often impossible to tell where one ended and the next began. Returning to Squaw Creek Refuge just before dark we expected to find many more geese resting on the water there, but discovered that the ones we had seen flying earlier had evidently carried on south without stopping. This mass migration was still going on, the long lines of geese passing over the refuge.

When we awoke next morning it was snowing and the ground was already carpeted with white. We humans could not have predicted the storm but the geese must have known something to have moved on south in such large numbers just ahead of the bad weather. On this snowy morning we too left Missouri.

Originally we had planned to follow the wild geese all the way to their wintering grounds along the Gulf of Mexico. But after losing four of the Creeps at De Soto we decided to revise our plans and leave the Central Flyway for a time so that our imprinted geese could fly again without fear of being shot by goose hunters. After leaving Squaw Creek we drove for four days to reach the Thunderhead Ranch, twenty miles east of Tucson, Arizona. Here we knew the Creeps would be able to fly in safety and they would even have their own private runway in the form of a disused airstrip. We all looked forward to warmer weather in the desert but Tucson greeted us with an all-time record snowfall of almost eight inches! However, it soon cleared and we enjoyed thawing out after many months in colder climates.

On reaching Arizona, in early December, Lee Lyon left to rejoin her family in California, having travelled with us for seven months. Les stayed on and our daughter Julie, then aged fifteen, flew over from her boarding school in Australia to join us for two months during her 'summer' Christmas vacation. Our one regret while making the snow goose film was that we had been unable to spend any time with her for almost a year, and while we were at the remote tundra camp even letters had been infrequent. Julie had longed to be with us on the tundra to help with raising the goslings, and now took over as goose-girl to the six-month-old Creeps.

10 We 'Fly' with the Creeps

It was a great relief to be able to allow the Creeps freedom again after the long road journey. Each day we took them in the Chevy station wagon to the dirt airstrip, but still allowed only a few to fly at a time until they became thoroughly familiar with their new surroundings. Biologists had earlier advised us that the imprinted geese were likely to become disorientated if we allowed them to fly in strange new areas as we travelled down the Central Flyway. We were grateful for the warning, but after months of travelling with the Creeps we are now convinced that if a goose wishes to do so it can always find its way back to the place from which it took off.

On the airstrip in Arizona we found that anything moving fast on the ground really excited the Creeps and made them eager to fly. This could be a car, a bicycle or even one of us running. By the end of December they were flying well and we felt the time had come to see if we could train them to fly alongside a moving car. Although we did not know how it would work out in practice, it was with this long-term project in mind that we had originally obtained permission to raise some stray goslings. The great migrations made by the snow geese are such an important part of their story that we felt we had somehow to get a camera as close to them in flight as possible. A few years earlier we had filmed a one-hour television special, *The World of the Beaver*, which owed much of its success to the fact that we had tried to show a beaver's-eye view of its world. For the beaver film we had therefore gone underwater. Now to achieve a similar effect with the snow geese we somehow had to get a camera actually among the flying birds. If our idea for filming the imprinted geese worked well we would have close-up scenes of flying geese that would be entirely different from film of wild geese that we had already taken from a helicopter.

We wondered if the Creeps would stay close to the travelling car or if it would only serve to get them airborne and then they would fly off on a course of their own choice. Maybe we would even have to keep some of the geese inside the car so that they could call to the flying ones and keep them close. We need not have worried, for the flying performances by the Creeps far exceeded our wildest hopes. Although it took a great deal of time and effort we soon had an efficient routine worked out. While Les drove the Chevy station wagon, Jen and I sat facing backwards on the lowered tailgate where the Creeps could see us clearly. Julie's job was to release the geese from a wire cage at the end of the airstrip just as the car began to move forward.

'Okay Les. Take it away,' I called when everyone was ready.

As the car leapt forward Julie opened the cage and the Creeps ran out, flapping excitedly.

'Creep. Creep, Creep. Come on Creeps,' Jen shouted, as we braced ourselves on the tailgate of the rapidly accelerating station wagon. Twenty yards behind us now, the Creeps took off and soon caught up with the speeding car. Now it was up to Les to keep the Chevy slightly ahead of the flying geese so that we stayed in the best position for photography.

'That's the way chaps!' I called, encouraging the geese. 'Come on. Closer you beauties. That's the way Creeps!'

Fortunately there was rarely anyone near to see or hear our strange activities, but our calls certainly helped to stimulate and encourage the flying geese. As the Creeps flew alongside they often called to us and we could watch their dark eyes following our every movement as we sped along at forty-five miles an hour or even faster. There was no doubt in our minds that they enjoyed having their foster parents 'flying' with them, for it was as if we were the leading geese in the formation with the Creeps taking up positions to the side and slightly behind. Often they flew so close that a wingtip brushed our heads or the cameras and at times we even reached up playfully to scratch a feathered belly suspended in flight just above our heads. Sitting on the lowered tailgate of a station wagon speeding over dusty ground is a thrilling sensation, even when holding on tightly with both hands. In spite of using a sturdy body-brace I found that holding a twenty-pound movie camera steadily enough to see through the viewfinder was almost impossible at first, but after weeks of practice we found ourselves riding that tailgate as if we had been doing it all our lives. There was a definite risk involved, but although we collected many bruises, neither of us fell off or was badly hurt. Our greatest anxiety was that one of the geese might collide with the car and be injured, but this never happened. Occasionally Les had to brake suddenly when some of the geese decided to land unexpectedly in front of the speeding car, but luckily we had no accidents.

At first, as a precaution, we allowed only five geese to fly at once but soon had all ten following the station wagon in formation. Nearing the end of the half-mile run Les shouted: 'STOPPING!' This warned us to brace ourselves for the sudden change in speed. Sometimes the Creeps landed alongside when the car came to a halt but more often they flew on for a few hundred yards before circling back to land. Down they came with feet lowered and bodies tilted so the tail was low, giving a few mighty flaps of their powerful wings as a final brake just before touching the ground. Immediately after landing, the geese always gave their feathers a quick shake to settle them. Then, as soon as we squatted down to 'talk' to them, they gathered around us jabbering loudly and seemingly very pleased with themselves. Three flights a day seemed enough for the Creeps. By the end of the flying sessions we were far more exhausted than they were, and completely covered with fine dust. The cameras needed cleaning each day, and as the fine dust penetrated the slightest crack, it often required hours of careful work to prepare the equipment for the next filming session.

We had worked with the Creeps for months now, first raising the orphaned goslings, then training them to be free-flying yet firmly attached to us. Now they had more than rewarded our efforts, both with their friendship and with their film-star performances. For the film taken of them flying alongside the speeding car is spectacular, making the viewer feel that he is seeing the geese through the eyes of one of the birds actually in the formation. Much of this filming was done with a 16 mm Eclair high-speed camera, which takes two hundred pictures every second: the normal speed for filming is twenty-four pictures each second. This high-speed filming resulted in some very interesting slow motion sequences, with eight seconds of viewing time for every second of live action. In close-up scenes the viewer is placed eye to eye with a flying goose, showing the tremendous surge forward with every downbeat of the powerful wings. The body actually moves up and down slightly with each wingbeat while the head remains level, causing the neck to undulate rhythmically. From watching the slow motion film we learned that, although all the geese flew at the same speed, some had more rapid wingbeats than others, in much the same way that people walking together have strides of different length, and are out of step. Many times when the Creeps flew together we heard the whack as the wings of two flying geese hit together in mid-air, but this happens so quickly that we have never detected it with the naked eye or on film, for it does not seem to throw the birds off balance at all. We are sure that this must be a common occurrence, especially during the massed take-offs. On seeing these massed flights in the Survival Anglia film *The Incredible Flight of the Snow Geese*, a major American aircraft company asked to borrow a copy in order to study the way the birds avoided mid-air collisions.

The Creeps soon learned to dodge the giant saguaro cactuses and the mesquite trees in the desert, but power lines are a definite hazard to all flying birds. Several times we saw our imprinted birds hit high wires and it amazed us that none were killed. On two separate occasions a Creep that had dropped like a stone after colliding with a wire turned out to be unharmed when examined immediately afterwards. Occasionally a slightly bloodied wing resulted, but fortunately no broken bones. One goose almost scalped itself hitting a wire but completely recovered. Another smashed the top of its upper mandible but within a week it had healed beautifully. There is no doubt that snow geese are tough birds, but we can't help wondering just how many wild birds are killed or maimed each year through collisions with power lines around the world.

When the Creeps were seven or eight months old the snows had white chests and bellies, with varying amounts of brownish-grey feathers remaining on the head and neck, as well as on the wings and upper back. Of the three blue geese, Slack Black – or Slacky for short – already had a completely white head and neck while one of the others had only just begun to show the first white feathers. The plumage of the third blue was somewhere in between these two, so this goose was christened Middy. Plumage development also prompted the naming of the two whitest snows as Whitey One and Whitey Two. It was inevitable that all the Creeps should eventually have names although we had resisted this at first. Slacky, Middy and Bub, were the three blues, while the snows were named Clark, Whitey One and Two, Pommy, Pete, Fuzz and Gerry. Gerry's name came about as the result of a collision with Julie when she was riding a bicycle along the airstrip one day. Excitedly the geese flew after her, landing in a group around the bicycle. She tried desperately to avoid them all but one was rolled over beneath the bike, emerging unharmed, but covered with dust and minus several tail feathers. Julie decided that as this goose was 'run down' it should be called Geritol, after the tonic.

To a stranger each of the Creeps looked the same, but we had no trouble telling them apart at this age, for each one seemed to be at a slightly different stage of plumage development. Some were larger and rangier than the others and although we had not sexed them we felt sure these were ganders. Some had slightly different shapes to their heads and bills, and we learnt to recognize several of them by the distinctive sounds they made.

Whenever the weather warmed to above 70° Fahrenheit we noticed that the Creeps were much less enthusiastic about flying. In fact on hot days one or two of the blue geese often dropped out during a long flight. Searching for the missing bird we usually found it panting in the shade of a bush, its chest heaving noticeably. With their darker plumage the blues absorb far more of the sun's heat than the snows; we only had to touch their feathers to find that the difference in temperature on the backs and wings of the two colour phases was astounding.

The Creeps now began to show occasional aggression towards strangers. Heads stretched forward and low to the ground, neck feathers vibrating, they would run forward and strike hard at the person's feet and ankles with their strong bills. If any one of us appeared in clothing that was strange to them we were likely to be subjected to the same treatment, unless we talked to them so they could recognize our voices. For instance, the Creeps had been used to seeing Jen in slacks and if she wore shorts or a dress they didn't seem to recognize her at first. This increased aggressiveness at least meant that the Creeps made marvellous watch 'dogs', for their loud and excited chattering always alerted us to a visitor's approach well in advance. Although they wanted to be with us whenever possible the Creeps had their own set of rules. If we lay on the ground it was fine for them to climb all over us, chewing on any part of our clothing or flesh that caught their fancy. But at the same time we were never supposed to touch them, and if we tried, no matter how gently, they usually shied away. On the rare occasions, however, when one of the Creeps became entangled in a piece of wire or had cactus spines in a foot, it instinctively knew that we were trying to help and stood perfectly still while we held it gently and sorted out the problem.

After they had had enough of flying the Creeps walked the half-mile back from the airstrip with us to the cabin where we stayed on the ranch. They cropped the meagre desert grasses as they walked, and spent the rest of the day in a large pen alongside the cabin. There they had a big pool where they could splash to their hearts' content. Though we would have preferred to leave the Creeps free all day, they often became a nuisance to people living nearby unless we watched them the whole time. If we left our cabin door open they came in to see us, but as it is impossible to house-train geese we soon had to herd them outside again. At night we walked the Creeps up a ramp into their trailer so that they would be safe from the numerous coyotes we saw in the area – they sometimes came at night to eat the chicken bones left outside the door of the cabin. We liked feeding the coyotes but we certainly did not want them eating the Creeps!

Because the film we were making was about snow geese, Fred and Canada Goose had to be kept out of the flying scenes. Although they were normally let loose during the day, we put Fred and Canada Goose in the pen while we were at the airstrip and let them out again when we returned. Fred's favourite form of exercise was to fly behind Julie or Les as they pedalled around the ranch on a bicycle. It was hardly a race, for Fred easily overtook the bicycle, often causing the rider to swerve suddenly when he landed on the track just ahead.

Fred's staple food continued to be the Miracle crumbles on which he had been raised, and which we still kept in the battered old paint can he had used at the McConnell River camp. He also developed a taste for quantities of cooked chicken; raisins; grapes; cheese, which had to be mild; spaghetti, which he swallowed in long strings like a worm; and green corn. He still refused to eat dry corn kernels – a favourite food of wild cranes in North America – but perhaps this was because he had the choice of too many other good things to eat. It was always amusing to watch Fred swallowing a grape whole. The grape had to be small to medium in size and he definitely preferred the sweet-tasting grapes like muscatels. As the fruit was swallowed whole its incredible progress could be followed all the way down: the round ball of the grape caused Fred's neck feathers to move noticeably as they were pushed outwards, with the bulge first travelling down the right side and

117

then moving around to the back of his neck, before finally being lost from sight under the right wing.

Whenever we were in suitable areas Fred was adept at catching and eating grasshoppers, beetles, worms, scorpions, spiders, lizards, frogs, fish and an occasional mouse. Slats Helm, the owner of the Thunderhead Ranch, really enjoyed Fred and brought him any mice he caught in the stables. Slats never tired of watching him swallow a mouse whole: it was even more dramatic than when he swallowed a grape. The more natural food Fred ate the happier we were too. He would follow us like a dog on our walks around the ranch and often surprised us by eating things we would not normally expect him to touch, such as the one-inch-long 'stink' bugs on the young mesquites. One afternoon he ate thirty-two of these large smelly shield bugs and was still looking for more. On another occasion during a late evening walk, Fred surprised us again. His eyesight is not good in the dark and even in the dim light he is inclined to trip over things we could see clearly. However, as we were walking together, Fred suddenly back-tracked about eight feet and swallowed a beetle. There is no way he could have seen it and we are convinced that it was some slight sound the beetle made that caused him to turn back and find it instantly.

On several occasions, while we were staying at the ranch in Arizona, Fred flew into Miss Harrington's yard – a neighbour and very good friend of ours from Canada. He had a marvellous time stalking and catching a goldfish in the pond: when this was discovered Miss Harrington, a great fan of Fred's, forgave him and continued to feed him titbits whenever he jangled the little bell on the outside of her gate to announce his presence. Fred regularly visited the Helms' yard, tearing up their carefully watered lawn with his strong bill, and probing in the soil with the tips slightly apart. We soon discovered that he was digging deep to catch the tasty worms that the Helms had actually bought and introduced to help make their lawn flourish. But once again he was forgiven. Fred often ate small amounts of green grass, always holding a piece crosswise in his bill before bending it in the middle so that it was swallowed with the two ends disappearing last.

By the time he was six months old, Fred's feathers had gradually moulted until he was a more uniform greyish colour, with red skin beginning to show through the feathers just above the base of his bill. During the next six months these feathers gradually disappeared until the front of his head was covered with a patch of red caruncles which brightened in colour and grew much larger whenever he became excited. Although most observers think Fred has a beautiful tail of long, curving, 'rooster-like' plumes, his real tail is quite short and, except when he flaps, it is completely hidden from view by the long flowing tertial wing feathers. Popular though the geese were with everyone, Fred was the star of the group, with a fan club stretching all the way from the Canadian Arctic to the Mexican border. After he had appeared in *The Incredible Flight of the Snow Geese* on television we even received fan mail addressed to Fred!

One day, after very heavy rainfall in the desert, Slats Helm was out on the main road, half a mile from the house, helping to tow a stalled car out of a flooded section of road outside the Thunderhead Ranch. Hearing the geese flying high overhead Slats looked up and saw his friend Fred flying with them. Without thinking, Slats called out:

'Hey, what are you doing up there, Fred?'

The moment he heard the familiar voice calling him, Fred put on his airbrakes, set his wings and came in for a fast landing alongside his human companion. The Creeps carried on, then circled back to land with Fred. In telling the story Slats does not repeat what the motorists said, but they must have been impressed by the sight of someone calling a flock of big birds out of the sky like that! Now that he had the Creeps and Fred on the road with him outside the ranch, Slats had problems, for the birds refused to fly home and he had to drive slowly ahead as he 'walked' them all back.

'Those old geese really take their time walking when they want to, don't they?' he said to us later. 'It took me over half an hour to bring them in after I'd called them down to me.'

Besides training the imprinted geese to fly behind the car, we had another idea for trying to film geese in flight and for over a year we had been making plans with model aircraft expert Larry Leonard of Los Angeles, California. By mounting a small 16 mm movie camera on a radio-controlled model plane, capable of flying at sixty miles an hour, we hoped to show a goose's-eye view of the wild geese flying. Because of our concern not to disturb the flocks of geese during the hunting season we had not tried out the model plane in the autumn. Now, in the winter, the most convenient place was at the Salton Sea National Wildlife Refuge in southern California.

A flock of perhaps two thousand snow geese, all white-phase birds, were feeding within one hundred yards of refuge headquarters when we arrived at the Salton Sea Refuge. We could hardly believe our good luck. This was an ideal situation for there was a surfaced road nearby for Larry to use as a runway for the model plane. After talking briefly to Bob Ellis, the Refuge Manager, we quickly readied the model plane and cameras for the first flight. But before we could begin filming, a high-flying commercial jet scared the entire flock, and after milling about in the air they landed at the far side of the field, almost a quarter of a mile away. Perhaps we were lucky that they did not fly out on to the Salton Sea, but now they were far more likely to be frightened by the model plane and would also have more warning of its approach. Even with a silencer, the small plane made a noise like a high-pitched motor mower as it warmed up on the roadway, although this sound alone did not panic the snow geese. They stopped feeding to look up but remained on the ground listening alertly, with heads held high. However, as soon as the red plane with the five-foot wingspan was airborne and within their view, the flock erupted, taking to the air *en masse* amid a chorus of calls. No doubt they considered it to be some strange kind of bird of prey that must be avoided at all cost. The plane soon overtook the geese, with the flock parting to let it through. Larry, expertly using two tiny levers to control its every move by radio, banked the little red plane and turned it back towards the birds. At all times he had to see the plane clearly in order to control its flight and the bright red colour of the upper surface, and white under the wings, helped him to keep track of its flying attitude. Whenever the plane was in position near the flying geese I was able to fire the camera by radio-remote-control. The model plane made four test flights that day and we obtained some interesting scenes, but the results were not as successful as we had hoped. Several months later we tried again, flying the plane near much larger flocks of snow and blue geese in South Dakota during the spring migration. Unfortunately, after half a dozen flights, the radio transmitter was soaked when the Land-Rover went into a ditch full of water. So we had to call a premature halt to the model plane's activities. The footage of this experiment was used in one of the half-hour SURVIVAL television films, but not in the Snow Goose Special.

After the goose-hunting season ended in Texas in mid-January we felt it was safe to travel with the Creeps to rejoin the wild snow geese. Although some flocks were scattered along the coastal marshes of the Gulf of Mexico in Texas and Louisiana, we found the main concentrations of snow and blue geese feeding on harvested rice fields west of

Des filming from the tailgate of a station wagon as the 'Creeps' fly in formation

Massed snow and blue geese fill the sky at Sand Lake Refuge
At De Soto Refuge the geese concentrate on an oxbow of the Missouri River

Clouding a full moon, snow geese rise above Sand Lake Refug

Fred loved to dry himself with a towel

Red-throated loon and chicks at the McConnell River

Houston. If the winter is mild some flocks of geese using the Central Flyway remain farther north in Kansas and Missouri, only flying all the way to the Gulf Coast in search of food and open water during severe weather. The flocks with a high proportion of blue geese that use the Mississippi Flyway in autumn normally winter on the coastal marshes and rice prairies of Louisiana, but in recent years some of these flocks have also been short-stopping on inland agricultural areas farther north.

Although the hunting on the wintering grounds may account for as much as seventy per cent of the geese killed annually on these two flyways, many hunters in Texas and Louisiana feel that the geese should be discouraged from lingering anywhere for long as they journey south in the autumn. So much money is spent by waterfowl hunters each year on equipment, accommodation and so on that political pressure has been used to try to make the northern refuges less attractive to the geese in the fall. But present day land-use practices on private agricultural lands have at least as much influence as the refuges on the behaviour of the migrating snow and blue geese.

While we were filming along the Central Flyway many newspapers in the various states reported at length on the goose hunting issue. What shocked us was that in the many articles that we read not once did a writer ever seem to care what might be the best for the future of the goose population. The main interest was to know which states would have the opportunity to shoot the most geese each year.

Many times we have been asked: 'What made you decide to study and film snow geese? Are they threatened with extinction?'

'No,' I am happy to reply. 'As things stand at present it is more of a success story with the snow geese. The flocks are carefully managed by the governments of Canada, the United States and Mexico and outside the controlled hunting seasons the geese are protected by an international treaty.'

The snow and blue geese are not as abundant as in earlier times, and probably never will be, yet they are today the most numerous species of wild goose in North America – and probably in the world. But this situation will only continue if they are managed wisely along the flyways and on the wintering areas, and are allowed to remain undisturbed on their tundra nesting grounds. There is talk of all kinds of development in the Canadian north and we fear for the future of at least some of the nesting colonies.

There are no federal or state refuges on the Texas rice prairies, although the geese soon learn to use private agricultural lands where there is no hunting. Often these private sanctuaries are actually supported by hunters who realize that unless the flocks of geese have somewhere to rest in peace they will be forced by hunting pressure to leave the vicinity entirely. This has been the philosophy of hunting outfitter Marvin Tyler, of Altair, Texas. As a result of giving the geese sanctuary on open water, where they can rest overnight, his hunters have been particularly successful. In one season, hunters shooting under the supervision of his professional guides have 'harvested' over 18,000 snow and blue geese. Marvin described to us the method of hunting he uses so successfully on the open Texas rice fields. As in South Dakota, the Texan hunters are in position before dawn and therefore before the geese are flying. Numerous pieces of white cloth are scattered amongst the rice stubble, with white clad hunters concealing their heads, guns and their dogs beneath pieces of white fabric. Snow geese flying overhead in the early morning light are attracted by a goose-caller and circle lower and lower to join what they think is a flock of geese on the ground, with the hunters waiting until the birds are well within range before suddenly showing themselves and opening fire.

After the end of the hunting season in Texas it was good to be able to let the Creeps fly near wild geese again. At first we stayed at a motel at Altair where we built a big pen for the geese under a large tree at the edge of a grassy field. Each day we were able to let them out for flying exercise and as before Fred usually flew with them. Because he would not wander far on his own, Fred was often left free to hunt for frogs and insects in a nearby swampy area, and he soon became well known to the people at the motel. One morning as we worked on correspondence in our motel room, with the door open so we could keep an eye on the Creeps and Fred, we heard the maids pause outside the neighbouring room.

'Hi ya Fred!' one called out in a strong southern accent. 'How y'all today Fray-ed?'

It was nice to know that we were not the only ones who talked to the birds!

11 *From Texas North in Spring*

When we went out filming in Texas we often took the Creeps with us, as they now travelled in comfort in the otherwise empty trailer. Because of the traffic hazards near main roads we only released them on private farmland or on wildlife refuges. All we had to do was to position a small wooden ramp and open the back door of the trailer. Fred was always first down the ramp, and invariably Clark led the Creeps, followed by Slacky – who was boss number two. We noticed no definite order for the rest of the Creeps, except that poor Middy was always last as she was at the bottom of the pecking order. Even her voice changed noticeably and she would wander about making the strangest goose sounds imaginable. It was as if she were telling the world that she admitted to being a 'hen-pecked' goose and would no one pick on her please!

Once they were all down the ramp, the Creeps flapped and stretched for a few minutes before taking to the air amid a chorus of calls. Now every time we allowed them to fly free there was a question at the back of our minds – would they want to come back to us or stay with their wild cousins? We had no hold over the Creeps apart from the family ties we had built up through the affection and protection we gave them. However, we certainly did nothing to encourage them to go. Biologists had warned us that band recoveries from birds shot by hunters showed that imprinted geese had a very poor survival chance when set free. At least if the Creeps did join the wild flocks now in the spring they would have several months in which to learn the ways of the wild birds before the hunting season. They would have a far better chance of survival than the Canada goslings that had been raised at the McConnell River a few years earlier and released at summer's end to fly south with no parents to guide them. The ones neck-banded JEN and DES had been among the few survivors of that group, but here we must add a sad note. During the autumn we heard from John Harwood, now back at the University of Western Ontario, that JEN and probably all four of her youngsters had been shot. Their bands had been sent in by a group of five hunters whose only regret was that they hadn't a sixth person along that day so they could have shot the one goose that escaped, for each hunter is allowed only one Canada goose. We assume that the escapee goose was DES, for Canada geese travel in small family parties rather than in huge flocks.

Whenever the Creeps flew, we called loudly to encourage them to stay in view but they gave us many anxious moments. We particularly recall one very windy day at Aransas National Wildlife Refuge when all ten geese flew higher and farther than ever before, completely disappearing from our sight two or three times while we called ourselves hoarse. From a greater height than we had seen him fly before with the Creeps, Fred poised himself to spill air past his firmly arched wings as he descended steeply with long legs trailing and already lowered. Nearing the ground he gave a few strong flaps to break the downward speed before landing beside us. Meanwhile the Creeps flew on, with the strong tail wind rapidly turning them into tiny specks near the horizon. I remember remarking to Jen and Les:

'This is a good way to develop ulcers!'

Eventually they did fly back, setting their wings and lowering their undercarriages before landing beside us on the grass. How we wished we could understand their 'goose talk' as they gathered around, jabbering excitedly, when we squatted on the ground beside them. The Creeps seemed so pleased with themselves and it was as if they wanted to tell us all about their wonderful flight and how exhilarating it felt to fly in such a strong wind. Although snow geese cannot talk as we do, they use a complicated set of sounds which are understood by all the birds of the same species.

One morning at Aransas we had let the Creeps loose while we filmed three quite friendly alligators in a pool near refuge headquarters. At first we were a little concerned lest the geese should decide they wanted to swim in the same pool. We need not have worried, for the Creeps instinctively feared anything strange, whether it was a potential predator

like an alligator or a harmless tortoise or skunk. They watched inquisitively as the alligators swam towards our side of the pool, but when one
came out on to the bank nearby, the Creeps retreated so that we were
between them and the alligator. Fred's curiosity prompted him to
advance a few feet in front of us for a better look, but he too was extremely cautious. When one alligator lunged forward on to the bank
to eat a piece of bread which had been left there, the Creeps took fright
and lifted into the air. After flying one circuit of the headquarters area,
they came in low over some bushes planning to land on the grassy slope
just beyond. But unseen by the flying birds, two collared peccaries, or
javelinas, had emerged from the undergrowth to feed quietly in the
open. Surprised by the noisy geese coming in to land almost on top of
them, the peccaries jumped with fright and ran off into the bushes.
This in turn scared the Creeps, who flew up again, causing a small group
of onlookers to burst into laughter at the comical scene. One man and
his wife had been travelling for over a month around the National Parks
and National Wildlife Refuges, and he remarked to us:

'Seeing the geese flying free like this, then setting their wings in the
strong wind to land right alongside us, is the most thrilling part of our
whole trip.'

It was certainly nice to hear him say it and although we do not know
his name, or even where he and his wife live, we do hope they were able
to see the finished film on television – and that it brought back many
happy memories of their holiday.

Later the same day the Creeps met with a larger group of peccaries,
but in cowardly fashion they flew to safety behind us each time the
animals came near. These peccaries are obviously conditioned to human
visitors around the picnic areas at Aransas, for one has only to open a
plastic bag of bread upwind, and they come running, expecting a hand-
out. But there is also plenty of natural food for the peccaries and we
were amused to see one chomping enthusiastically on a meal of green
stinging nettles.

To escape the persistent peccaries the Creeps finally retreated into the
nearby waters of the Gulf of Mexico for a swim. This was their first
experience of salt water, and they revelled in it. However, evidently
remembering the alligators, they cautiously stayed in the shallows
close to shore while they bathed and splashed, often turning completely on their backs with their heads underwater. Fred took advantage
of this opportunity and had his first real bath in weeks, a procedure he

usually indulges in when he thinks nobody is looking. Wading out into water up to his belly he dips his bill in a few times, flicking the surface repeatedly as if testing the quality or temperature. Satisfied, his head and neck go underwater several times, an activity that is accompanied by much splashing. He then rolls first on to one side, then on to the other, in order to wet his back completely before standing up to beat the water briefly with his wings, sending spray in all directions. A few quick flaps while standing in the shallows shakes off any excess water, and on wading ashore the long preening process begins. Although Fred industriously oils his feathers in much the same manner as the geese, he is not equipped to do such a thorough job of waterproofing. Following a bath, his feathers become quite soggy and bedraggled and it is often an hour before he is capable of flying again. While the geese bath at every opportunity, with Fred it is at best only a weekly occurrence, though he spends hours preening every day.

We saw much of interest at Aransas, including whooping cranes in the distance, and would gladly have spent more time there filming the variety of birds, mammals and reptiles, but our main project for Survival Anglia Limited was still the snow goose story. A few miles outside the refuge there was a flock of three or four thousand snow and blue geese feeding in a field very close to farm equipment and buildings. It was extremely difficult for us to show on film that the wild flocks of snow geese had indeed reached their southern wintering grounds. We drove for hundreds of miles and saw dozens of pumping oil wells in agricultural areas, yet we were never able to find large flocks of geese nearby to show they really were in Texas. But we did manage to film snow geese feeding side by side with Texas cattle, and from the air we were able to show flocks along the coastal flats with the waters of the Gulf of Mexico beyond.

Although some flocks were scattered along the coastal refuges most of the snow geese that we saw in Texas were concentrated on harvested rice fields in the vicinity of Eagle Lake. There is a surprising similarity in the areas used by the snow geese along their flyways. The flat rice prairies of Texas closely resemble the prairie corn fields in the Dakotas, which in turn resembled the level fields of rice and other grain we previously saw used by the Pacific Flyway birds in California. These regions in the south bear a likeness to the wide open spaces of the tundra nesting grounds.

During the winter of 1971–72 relatively few snow and blue geese utilized the Louisiana coastal marshes. The Sabine National Wild-life Refuge is situated in prime goose habitat, but their peak snow goose population that season occurred in February – with only ten thousand birds. So there was little snow goose filming we could do in Louisiana.

For a number of months I had been corresponding with Johnny Lynch, a well-known goose and crane expert working with the U.S. Fish and Wildlife Service at Lafayette, Louisiana. So before heading north we were able to visit his home and Orchid Gardens and talk about snow geese with Johnny. From him we learned about their ways in his home state, where he has been conducting winter goose surveys for over twenty years. At his home we were treated to a veritable feast of a local delicacy, freshwater crawfish, or crawdads as they are called here. John's wife Zoë cooked forty pounds of these five-inch crustaceans, but only the tails are eaten and the dozen people present made short work of the feast.

We returned to the Eagle Lake area of Texas to wait with the big con-centration of snow and blue geese until they started their long spring migration north. This enabled us to see and film the fascinating spring display of the Attwater's prairie chicken, an endangered species. Near Eagle Lake there is a World Wildlife Fund Sanctuary, where natural prairie grasses are maintained especially for the benefit of this threatened species. For our first visit to the refuge we arranged to meet the spritely 80-year-old warden, Tom Waddell, before dawn outside the Eagle Lake post office building. With the Land-Rover closely following his pick-up truck, we were soon inside the locked gate of the reserve well ahead of sunrise. When we finally stopped, switching off both the headlights and the engine, an incredible sound seemed to envelop us – the booming of the Attwater's prairie chickens! How could a bird make such a sound? It seemed unreal, utterly impossible to describe, rising and falling and repeated over and over by perhaps a dozen males, each trying to outdo the others in both volume and variation of calls. A person can, with mouth closed, hum the eerie *oooo – oooo – ooooo* sounds in a similar wavering manner to the male prairie chickens, but only after becoming familiar with the strange booming hum that they make.

As it grew lighter we made out the shadowy shapes of the males, strutting about the ancestral booming ground, with drooping wings held out from the body and primary feathers spread. Their general colour is brownish-fawn, barred rather than speckled, but it is the bare, colourful orange pouches, one on either side of each male's throat, that immediately claim one's attention. These are distended, balloon-like, to create the booming hum as the male advances slowly with head lowered, nape feathers held forward like a crest, and the tail cocked at a jaunty angle. Two males face each other aggressively, leaping into the air like fighting cocks to slash at one another with sharp spurs. Another male is busy doing a little solo dance, leaping into the air and making fowl-like clucking sounds, as his wings rapidly beat the air and his feet are drawn in close to the body. Once back on the ground he struts jauntily forward with the bright orange gulah pouches distended as the booming sound begins again.

During our first visit we did not notice any female Attwater's prairie chickens on the booming ground, but on later visits to the reserve with Mike Mihalo, a graduate student studying these birds in detail, we became more familiar with the activities of the prairie chickens. Once we counted twenty-two displaying males, but only four females seemed impressed enough to put in an appearance; perhaps others were already sitting on a full clutch of eggs. Slightly smaller in size than the males and similarly marked, except for the lack of colourful throat pouches, the females plucked at short green grass without paying attention to the strutting cocks. Although the females seemed little concerned with the proceedings, the males immediately intensified their strutting and the booming calls quickened after the females appeared. Finally one female seemed to favour a male, half crouching in front of him with quivering wings. He advanced towards her, booming steadily, only to turn away at the crucial moment to fight with another male. Still the brown-barred female waited quivering, and the male finally returned to her. We fully expected to see them mating, but nothing happened, as the male left once more to attack another would-be rival; it seemed as if aggressive behaviour was far more important to him than taking care of the waiting female. Who could blame her for finally losing interest and walking off the booming ground.

Not long after sunrise there is a lessening of activity, with some males flying off into the surrounding grassland. Other males continue boom-ing, but half an hour later the short grass arena is deserted. Late in the afternoon the male prairie chickens usually, but not always, return for another session on the booming ground, and the eerie humming sounds continue until sunset. For three months, from February to April, the prairie grasslands are enlivened each dawn and most evenings by the spectacular performances of the male prairie chickens. We felt privileged to share their world even for just a few weeks.

In early March colourful displays of Indian paintbrush, bluebonnets and phlox bordered the roadways in southern Texas and with the blooming of the wild flowers, the flocks of wild geese became restless. On the afternoon of 10 March the skies began to clear after several days of overcast weather and northerly winds. That evening, just after dark, we heard the unmistakable gabbling of wild snow geese in flight high above our motel. We rushed outside and listened, but the geese were not visible in the night sky. Almost non-stop for half an hour the sounds continued, as flock after flock of geese headed northwards on the first stage of the long migration to their tundra nesting grounds as much as three thousand miles away. While the weather conditions remained favourable for the next few days, flights of geese headed north in long Vs, with the main exodus beginning around dusk each day and continuing for over an hour. Some flights contained as many as six hundred geese but the average was around two hundred birds.

The Texas rice farmers are always happy to see the geese – and the sandhill cranes – leave in the spring, for it is then safe to plant new crops without the worry of depredations by these birds. For the geese too it is safer to leave before the rice is planted, as the seeds are often treated with pesticides to prevent water weevils consuming the germinating grain. If enough of this treated grain is consumed it can be lethal to waterfowl, and in years when some geese remain until late March or sometimes well into April, there is mortality among the flocks.

Now that the wild flocks were heading north it was time for us to follow with the trailer-borne Creeps, for we planned to film one complete twelve-month cycle in the life of the snow geese. We knew that their first major stop would be at Squaw Creek Refuge in Missouri, some 750 goose-miles north. Before leaving Texas we telephoned Harold Burgess and learnt that there had been a large influx of snow and blue geese at Squaw Creek on 11 March, about twenty-four hours after we had heard the first large flocks leaving Texas. From our experience with the Creeps flying behind the speeding car we knew that geese have no trouble in flying at forty-five or fifty miles an hour and with favourable tail winds can travel at much greater ground speeds.

Reaching Squaw Creek on 24 March we found only twenty thousand snow and blue geese using the refuge, for in spring there is no hunting pressure and the birds can safely spread out on farmlands to feed on grain left over from the autumn harvest. Farmers are more than happy for the geese to clean up the waste grain as this prevents unwanted volunteer crops from springing up. With the thawing of winter snows there are many ponds scattered about the agricultural lands which the geese use in spring, so water is not a problem.

Although we no longer had to worry about hunting when allowing the Creeps to fly, we decided to keep them grounded during the few days we stayed at Squaw Creek. Around refuge headquarters huge trees form an almost solid canopy overhead. Our geese were not familiar with this sort of hazard and we were concerned that if they managed to take off through the gap in the trees they might have great difficulty landing again. Our fears were confirmed one day when Fred ran flapping near four of the Creeps that were outside the pen feeding on grass. This activity excited the geese into wanting to fly, but only Clark managed to gain sufficient altitude to clear the tall trees. Once aloft he was unable to see us but sensibly flew back and forth calling loudly. We answered and all the other Creeps called repeatedly, but Clark obviously had no idea how to go about landing among the dense trees. Finally Jen ran a short distance down the road and called to him from a small clearing. Once, twice, three times Clark flew above her calling, then judged his approach carefully and made a perfect landing. One very relieved goose was back on terra firma and we could all relax. After his frightening experience, Clark did not utter a sound until a few minutes after he had rejoined the other Creeps.

Many of the wild geese had already moved farther north so we followed by road, hoping to overtake the vast flocks in the Dakotas. Pausing briefly at De Soto Refuge we arrived at Sand Lake Refuge on 27 March to find that we had not quite raced the first of the wild geese. Due to an early thaw the snow had disappeared and although the waters of Sand Lake itself were still frozen, many of the shallow sloughs in the neighbouring countryside were free of ice. A vanguard of about twenty thousand snow and blue geese had recently arrived, preceded by many thousands of ducks, mostly mallards and pintails.

Driving farther north from Sand Lake we saw no geese from the roads, and soon found ourselves in areas of southern North Dakota where snowy fields and ice-covered ponds prevented the geese from continuing north. The spring migration of the snow geese is quite different from their almost leisurely southward journey later in the year, for there is a great urgency to reach their northern breeding grounds as quickly as possible. As soon as rivers and shallow ponds thaw and some of the agricultural lands become free of snow, the geese push northwards in short hops. With the absence of hunting pressures in spring they can land freely on farmlands where food is more plentiful than on the chain of refuges where they concentrated so heavily during their migration the previous autumn. While the geese very rarely remain in any one area for more than a few days in the spring, it is possible at this time to see even greater concentrations than at any other season. This is mainly due to the urgency of the spring movements and also to the presence of some of the geese that flew non-stop from James Bay to the Gulf Coast in the autumn. This would account for the higher percentage of blue geese that we noticed among the spring flocks on the Central Flyway. On the southbound migration blue geese made up about twenty per cent of each flock, but now we noticed that some concentrations contained almost as many blues as snows.

To find out more accurately where the flocks of geese were, we chartered a light aircraft on 1 April and flew north from Aberdeen, South Dakota. From the air we followed the James River – as the geese also do – and could clearly see that the flocks had pressed on to the farthest point where they could find open water. Just inside North Dakota the advance flocks waited for warmer weather to hasten the thaw before flying on to the area around Devil's Lake. We went on into North Dakota ahead of the geese, and noticed that at first only scattered patches of snow remained on the farmlands below, certainly not enough to prevent waterfowl from feeding. It is said that in spring the geese bank up against the snow in their haste to push northwards, but from what we saw from the air it seems that the availability of open water is the main factor governing their movements. We actually flew well over fifty miles north of the most advanced geese before the ground below us was completely snow-covered. In some years a sudden spring storm blankets the land with snow and freezes the ponds hard once more, forcing the waterfowl to retreat temporarily many miles to the south in search of open water and food. Flying over the thawing farmlands we were fascinated by the patterns on the fields passing below, many appearing to be made up of black and white stripes, the white lines being formed by snow still lying in the long straight furrows. Small frozen ponds dotted the landscape – the potholes that make it possible for the northern prairies to become a veritable duck factory during each year's nesting season. We flew as far north as Devil's Lake and found the lake itself still completely frozen, with snow blanketing the surrounding countryside.

During our aerial survey we noticed that the largest flocks of snow and blue geese were concentrated just south of Sand Lake, on agricultural land and on a series of sloughs, or marshy areas. These great flocks were not much more than fifty miles to the south of the smaller vanguard flocks of geese we had seen from the air. For our aerial photography we had chosen a clear and relatively mild day, with the temperature several degrees above freezing. Later that afternoon, after driving back to Sand Lake Refuge, we suddenly heard the distant clamour of many snow geese on the wing. Looking to the south we saw long diagonal lines and Vs of geese heading our way and these soon stretched from the horizon to high above our heads. Using binoculars we could see dense flocks taking to the air several miles to the south, gaining altitude as they grouped into their typical formations. These formations are constantly changing as the flocks travel and no one goose remains in the lead for long. A diagonal line of migrating geese will suddenly undulate and bend in the middle to form a V, with the lead goose now at one end. Conversely one arm of a V may drop back slightly and move over to join the opposite side to form the one line. But no matter what their exact formation the geese always position themselves so that each bird is behind and slightly to the side of the goose immediately ahead. In this way every goose in the flock has a clear view forward and yet each one is close enough to the bird ahead to benefit from flying in its slipstream. It would be difficult to experience a more moving and thrilling sight in nature than the seemingly endless gaggles of talkative northbound geese outlined against the deep blue sky overhead, with their wings making clearly audible swishing sounds in the calm air of a late afternoon.

Judging from our aerial observations earlier in the day, these geese would not be flying more than thirty or forty miles farther north, although if the milder weather continued they would soon be pushing well into the Devil's Lake area of North Dakota. So we followed in the Land-Rover to see where the geese would be stopping. For over an hour the flights kept coming in the evening sunshine, perhaps fifty thousand geese or even more, and although they seemed to arrive in three or four main waves there were birds in the air all the time. We watched as some of the geese decided to settle for the night and joined others on a slough to the north-west of the refuge. It was interesting to see the momentary indecisions within a flock as some of the birds spiralled downwards towards the water while the remainder decided to continue northwards. The geese do not maintain the same grouping within the flocks, as family groups frequently break away to join up with other flying geese, separating and regrouping almost constantly on these short spring migration hops. For several minutes after the last flights had disappeared from our sight their calls came clearly back to us as we stood marvelling at the spectacle we had just witnessed.

12 *In the Dakotas and Manitoba*

It had not taken us long to reach the conclusion that snow geese and humans have entirely different ideas about what constitutes agreeable weather conditions. Every time the temperature warmed enough for us to enjoy being outdoors the geese would be off and we would have to follow. At Sand Lake Refuge we once again welcomed the use of the Bird House, as the spring weather is even less pleasant than that of autumn.

The Creeps were given the use of the two large enclosures. But now with no hunting going on they could all be free much of the time and they had their first real flying exercise since leaving Texas. In early April we selected a few long stretches of dirt roads inside the refuge where we could continue to film the flying Creeps from the tailgate of the speeding car, but now we had typical snow goose habitat in the background as they flew above cropland, lake and marsh. Wild geese often flew nearby, but the Creeps took little notice of them and always returned to land alongside us, although one day Middy had an interesting meeting with a wild goose. As mentioned earlier, poor Middy seemed to have fallen to the bottom of the pecking order in our little flock and as a result sometimes lacked the stamina of the other Creeps when flying. Several times this blue goose dropped out of a flight and we would later find her patiently waiting to be picked up, but always calling loudly so that we had no trouble in finding her. Once when we drove back along the road to collect our dropout we found a lone snow goose had landed alongside Middy, obviously thinking it had found a mate. As we approached, both geese took off together and we held our breaths anxiously until Middy flew back to land near us, while the puzzled wild goose circled overhead briefly before flying off.

The results of film taken earlier with the special high-speed Eclair camera had not been consistent; many rolls had been ruined by an intermittent fault, judder, while other rolls were perfect. The good rolls were so good that we persisted with this highly specialized camera, for if we sent it to the factory in France for a complete overhaul we would never have it back in time to use again on the snow goose filming. All our exposed movie film was shipped unprocessed to Survival Anglia Limited in England, and as soon as we received reports that this camera was playing up we had it checked in the States, but the problem persisted.

On one flight at Sand Lake the Creeps performed perfectly, flying in a classic formation close to the car as we sped along the curving road for over a mile. I was able to shoot a whole 400-foot roll of slow motion film without once taking my finger off the firing button, with marsh, lake and wild snow geese showing in the background. When Les brought the Chevy to a halt at the end of the run I was elated at the way the Creeps had performed and how everything had seemed to fit perfectly together.

I remarked to Jen and Les: 'If the camera played up during that roll it has to be the best roll of rejects I've ever taken!'

We would like to be able to report that we had a whole roll of fantastic, continuous, slow motion flying scenes of the Creeps, but unfortunately the camera DID play up and the whole of the roll was jittery and absolutely useless! Of course it was many weeks before we heard about the film, but this is one of the many uncertainties one has to cope with when filming in remote areas. We often say that if wildlife filming were easy, many more people would be doing it. It is certainly not an occupation for anyone who is easily frustrated or discouraged.

We look back on those April days along the shore of Sand Lake as some of the most enjoyable and memorable that we were privileged to spend with our imprinted geese. After each flight we let the Creeps enjoy themselves on the lake shore, where they ate grass and used their powerful bills to dig up water plants and eat the tender portions near the base. The geese do this with heads submerged and eyes open, first paddling with their feet in the shallows to loosen bottom material before tipping up to work underwater with their bills. By April the

Creeps' legs and feet were becoming much redder as a pale outer layer of scaly skin dried and flaked off. A few had almost attained their adult plumage, and they were looking extremely handsome. At first we left Fred and Canada Goose in the enclosure at headquarters while we filmed the Creeps flying, but they did not think much of this idea. In fact Canada Goose had fits of sulking, often refusing to come near us when we returned from one of these outings. So we built a small pen by the lake shore where the two of them could be kept out of the way while we were actually filming, and could later join the Creeps at the edge of the lake. Harmless garter snakes were common in the grass along the lake shore, but more than once we noticed a goose actually tread on a snake without ever seeing it. Neither they nor Fred were really frightened by these small snakes although they avoided them whenever they saw one.

To make sure that Fred had some exercise we sometimes made him fly behind the car during the two-mile trip back to headquarters, but we had to be extremely careful. If we had to slow down to negotiate a sharp bend in the road Fred often flew ahead to land just in front of the car, for he seemed to think that being forced to fly home while the geese rode was beneath his dignity. Cranes are very graceful in flight and this was a wonderful opportunity to study Fred's flying action at close range. With neck stretched forward and long legs trailing to form one straight line from bill to toes as he flew, his large wings made slow, effortless downbeats. With a flick of the wings as they began to lift, the upbeat was much more rapid. In the air Fred was a joy to watch but his landings were often far from graceful. Perhaps the difficulty cranes sometimes have in making perfect landings has to do with their exceptionally long legs. Whatever the reason, Fred often skidded on gravel or tripped clumsily over grass tussocks and his feelings seemed hurt if he made one of these awkward landings in our presence.

Although we had seen many thousands of snow geese moving north on 1 April, others continued to arrive in the region of Sand Lake from farther south. During the first week in April a peak population of about 200,000 geese was estimated to be in the area. The weather turned bitterly cold again at this time with the temperature not rising above freezing for a few days. In spite of the cold, and a light snowfall, we were out filming the massed geese whenever possible, obtaining scenes of them feeding among the cut corn stalks. Such scenes had been impossible to film during the autumn when the wild geese were much

'Clark Gable' – boss of the 'Creeps'

Des uses a high-speed camera, taking 200 pictures per second

Spring mating display of western grebes at Sand Lake Refuge

more wary because of the hunting season. There was constant noise and movement, with a flock of several thousand lifting off with a roar every few minutes to resettle a short distance away. Before the end of the first week in April the weather suddenly warmed up and by mid-April there were just a few scattered flocks of snow geese in the vicinity of Sand Lake. A few small groups lingered there until early May but these were probably birds that for one reason or another would not be nesting this year and therefore they were in no hurry to reach the tundra nesting grounds.

With no hunting to help keep the geese out of their new crops in spring, some farmers use noisy carbide guns to try to scare the lingering flocks from their fields. In California we had seen a different method used to frighten geese. Fields were sparsely dotted with posts to which squares of foil were attached that shone brightly and rattled noisily whenever moved by the wind.

We had planned to complete our full year of filming the snow geese by following them as far as Canada where up to a million of them congregate in spring on the prairie farmlands of southern Manitoba, the last suitable resting and feeding place before they fly northwards over forested regions to the shores of Hudson Bay. At the end of March we drove to Winnipeg on a reconnaissance trip to gather information about the spring movements of snow and blue geese in Manitoba. Eugene Bossenmaier, a biologist with the Manitoba Department of Mines and Natural Resources, was most helpful. We learned that beginning in mid-April the geese gather on agricultural lands to the south-west of the city of Winnipeg, capital of the province of Manitoba. Here they feed and rest, usually until the second week in May. Mating is frequently observed among the geese here; in years when adverse weather conditions delay their departure for the north, farmers in the area report finding quite a few goose eggs dropped haphazardly in their fields. When the snow geese finally leave for their northern nesting grounds, many of the flocks pass right over the city of Winnipeg, causing residents to pause in their daily routine to gaze aloft.

Although the migrating geese are a welcome sign that summer is not too far away, they present a hazard to aircraft and in 1969 a civil airliner was seriously damaged when it struck a flock of snow geese near Winnipeg. Since 1970 the spring migrations of snow geese in southern Manitoba have been monitored by the Associate Committee of Bird Hazards to Aircraft, of the National Research Council of Canada. Hans Blokpoel and his researchers have each year used visual as well as radar techniques during early May to trace the size and speed of the migrating flocks, which usually fly over southern Manitoba at altitudes around 1,000 or 1,500 feet. Using radar equipment at Winnipeg International Airport, Hans Blokpoel has established that the majority of the snow geese may leave the prairies in one large wave with almost continuous departures over a one or two day period. In some years there may be more than one significant mass movement, but usually within the space of a week virtually all the geese have taken off on the five to six hundred mile flight to the shores of Hudson Bay. It appears that the snow geese migrate from southern Manitoba on a broad front one hundred miles or more wide. Further to the south, while making shorter flights, they follow rivers and valleys as they push north, but on leaving the prairies the geese do not appear to follow any pattern of geographical features. If they have been delayed on the prairies by bad weather, the start of the mass migration may take place before the weather has become really favourable, so eager are the geese to push on. This was the case in 1970 when the mass exodus took place between 15 and 17 May. Unfortunately for us, in 1972 the main exodus was over by the morning of 11 May, and we arrived in Winnipeg that very afternoon hoping to film it!

'You should have been here this morning,' said the manager of the motel where we stayed. 'Those geese were flying right overhead.'

On talking to Hans Blokpoel I learnt that this was the third large movement of geese and that an even bigger flight had taken place a few days earlier, but most of the geese had flown a little to the west of Winnipeg this year. The first wave of geese had actually left on 4 May and large numbers had been reported arriving at York Factory, on the south-west shore of Hudson Bay, on 5 May.

After all our earlier planning it was a disappointment to miss filming the migration over Winnipeg. This is the only big city over which the lesser snows pass in large numbers during their travels, but we had been delayed by extra filming in the Dakotas and hadn't been prepared for the geese to depart slightly ahead of their normal schedule.

From the very beginning of our snow goose project, our idea had been to make a very detailed one-hour film about the life of wild lesser snow geese, keeping the imprinted geese completely out of the picture except for the close-up scenes of them in flight. When we trained the Creeps to fly behind the car in January, I felt that the 'behind the scenes' story of how these flying shots were made was too interesting not to be filmed. For a long time Survival Anglia Limited had been thinking of making a television programme showing how their wildlife cameramen work in the field, and I felt a short coverage of our activities would probably be useful.

Everywhere we travelled, people were fascinated by our free-flying imprinted geese and through our letters, people working at Anglia's London office also became interested. The scenes showing how we had filmed the Creeps flying made everyone wish we had taken more film telling the story of the imprinted geese and their experiences while travelling with us down the Central Flyway. It was easy to be wise after the event; if we could have shown the migration of our imprinted geese paralleling that of the wild geese, the resulting film would have a wider popular appeal than a film solely about wild geese, and would be assured of a network showing in the United States. It was already well into April when letters dealing with all this began flying back and forth between Colin Willock, at Anglia in London, and ourselves at Sand Lake, South Dakota. The snow goose project had already been an extremely time-consuming and expensive operation. The only way that we could do the extra filming showing the imprinting of goslings, with a minimum of additional time and expense, was to return to the tundra in June. This time we could camp near a smaller snow goose colony only thirty miles east of Churchill and film the imprinting of a new family of orphaned goslings which would, in effect, be doubling for the Creeps as youngsters.

Already it was late April when these plans began to take shape, but we did not need to be on the snow goose nesting grounds until mid-June, in time for the hatching of the goslings. We could therefore stay around Sand Lake to film the wildlife of the prairies during the spring. We would also need the intervening weeks to film more of the activities of our free-flying Creeps. Inevitably this involved occasionally getting on the wrong side of the camera ourselves, which none of us enjoyed, and Les had to be given a crash course in using an Arriflex to film some of these scenes of us both with the Creeps. Julie was also with us at Sand Lake during her May holiday from boarding school in Australia, and she helped in every way possible.

By the time these plans began to take shape virtually all the wild snow geese had moved a few hundred miles north of our temporary base at Sand Lake, and we particularly wanted scenes showing the Creeps flying near the wild flocks. This was something that had occurred many times in the spring, the imprinted geese always landing with us rather than with the wild geese. There were no wild snow geese remaining on the refuge, but Lyle Schoonover told us of a flock of a few thousand that was lingering on a marshy area several miles to the east. These birds were within two hundred yards of a main road but luckily there was little traffic and we felt it would be safe to let the Creeps use the road as their runway.

The Creeps took off from the grassy verge beside the main road, and aided by a strong wind, quickly gained altitude, first making two wide sweeps to the west, away from the wild geese. As they passed high overhead and flew to the north we saw the little group split. Seven of the Creeps obviously wanted to fly over to visit the wild geese, while the other three opted to return to us. Anxiously we watched and called. Soon four more split from the truants to turn towards us and finally the adventurous three also decided to rejoin their friends. But our worries were not yet over for after passing overhead all ten Creeps headed back towards the wild geese, some of them actually setting their wings to make a landing approach before heeding the calls of the other Creeps and ourselves and returning to land beside the vehicles. We had all been sure that at least some of the Creeps were going to land out in the marsh with the wild flock and had envisaged a long vigil by the roadside while waiting for them to return to us, if indeed they chose to do so.

Fortunately there had been little traffic on the road while the Creeps were flying and only two cars stopped to watch. One was driven by a farmer who evidently went home and described what he had seen to his family. I heard the outcome of this the next day when I visited the little bank in nearby Columbia and mentioned to Martin Weismantel, the friendly manager, what we were doing in the area.

'So that's a true story I heard this morning,' he said.

'What exactly did you hear?' I asked.

'Well, a customer of ours told his sister that he'd watched some people let a bunch of snow geese out of a trailer near a large flock of wild geese. After a while the tame geese flew over to the wild geese and circled around before flying back to land near the people. He told me his sister was quite prepared to believe all this, but when he went on to say the people just walked their geese up some steps into the trailer, closed the door and drove off, this was too much, and she was sure that he was pulling her leg!'

If the Creeps had joined a wild flock before this time it would have been completely of their own choice and we would have wished them well. Now, however, we really needed the imprinted geese to stay a while longer for filming. So after our most recent experience we decided not to fly them near wild geese again until we had finished filming them, just in case.

Back at Sand Lake Refuge, which the Creeps had almost decided was their home, we borrowed a canoe so that we could look for nesting birds among the cattails and at the same time take the Creeps on to the lake for outings. They loved to swim alongside as we paddled, and every so often took off to fly in sweeping circuits around us before landing again on the water nearby. If they became bored with swimming they thought nothing of flapping their way up into the canoe, showering us with water as they did so, and often wetting the cameras as well. Looking very pleased with themselves they stood happily in the canoe watching the passing scenery until we reached shore again. Fred was usually left on shore to amuse himself looking for insects while we went canoeing, but often, deciding he must be missing something, he flew out to join us, landing gracefully on the water nearby. Although not equipped with webbed feet for speedy progress, Fred's long legs paddled hard while he held his folded wings carefully out of the water and tried valiantly, but not too successfully, to keep up with the swimming Creeps. Sometimes we felt sorry for Fred and lifted him into the canoe for a free ride, but at other times he became bored with the whole business and flew back to shore. We were surprised that he could take off quite easily from deep water. Wild cranes spend a great deal of their time in shallow water but they would normally have no need to go out of their depth.

When we all returned to shore the Creeps usually grouped along the bank preening and resting. Fred normally followed suit, but one day he became very excited at finding Les's socks which had been left on shore beside a pair of boots. Picking up a sock in his bill, Fred spent the next half hour rubbing it all over his feet, legs, underside, wings and back, all the time purring excitedly. Many times he dropped the sock only to pick it up again and eagerly carry on with the wiping motions. It was almost as if he were towelling off after his swim, but later we found he sometimes reacted in the same way if given a small piece of wet cloth even when his feathers were dry. If we washed the car while Fred was around he became almost frantic in his efforts to take the wet rag away from us. Usually we tore off a six-inch square of cloth for him to use and he would spend half an hour or more standing beside the bucket of water, alternately dipping the cloth in the water and wiping it all over his body, wings and legs. We are unable to explain this behaviour of Fred's although in the wild, pairs of sandhill cranes are said to 'paint' themselves in a similar fashion as part of their courtship ritual, staining their feathers various shades of rusty brown from mineral deposits in the marshes. Fred's towelling-off act was so amusing to watch that I decided to film it as a short sequence, even though we had been specifically asked to keep Fred out of all scenes in the snow goose film. Whether in the flesh or on film, Fred has a way of captivating everyone's attention, but nobody was more surprised than ourselves when this sequence was used in the one-hour television special, *The Incredible Flight of the Snow Geese*.

In late May, Fred's purring sounds deepened noticeably and on 6 June he surprised us, and himself, by uttering his first adult crane call when he became excited by the Creeps flying nearby. He had never heard another crane calling but it seems that the basic calls of all species of birds are built-in and do not need to be learned.

The story of crossing the Canadian-United States border the previous autumn with our family of geese had amused all who heard about it. So for filming purposes we decided to restage this incident at the Pembina border post with the full co-operation of all the officials concerned. But this time, to make a more interesting film sequence, we stretched the original story a little and released all ten Creeps at once. While the Veterinary, Immigration and Customs officers looked on, the geese walked slowly down the trailer ramp on to the roadway. After a few quick flaps to stretch their wings they took off into a southerly wind. Circling back to pass over the border post they headed north into Canada with a tail wind speeding them on their way. Half a mile or more beyond the Canadian border the little flock flew back and forth low over the open farmland for a few minutes, then headed north again until they were almost out of sight. Finally they turned back and were soon making a circuit high above us, but now we could count only nine geese, for evidently one of the snows had stopped off somewhere inside Canada, probably where we had seen them flying low over a field. However we still had problems with the nine airborne Creeps, for instead of landing as we had expected, they merely circled high above us before heading north again. This time they flew completely out of sight, convincing the onlookers that they had merely come back to give us a farewell salute before flying off for good to join the wild geese farther north.

'You'll never see them again, will you?' asked one Customs officer.

'It's entirely up to the geese when they're flying,' I replied. 'But so far they have always come back to us.'

Anxiously Jen scanned the northern skyline through binoculars as she searched for the truants. Several minutes passed before she spotted them.

'I can see them now,' she called. 'They're flying back and forth way to the north. Now they're heading this way. You should be able to see them any moment.'

As they flew nearer we were relieved to see that there were still nine geese flying. Before landing they made several wide circuits of the border post, at times passing out of sight behind large trees a few hundred yards to the east. At last they touched down on the grass where we waited beside the road – but now we had only eight geese! Evidently Middy, the blue goose, had tired and landed behind the trees during one of the final circuits. It was a great relief to have most of the Creeps back safely, and the border officials were very impressed, as they had been convinced we would never see any of our geese again!

Although the Canadian border officials knew about our filming activities we felt more than a little sheepish at the thought of having to ask for permission to go into Canada to search for our lost snow goose, but it seemed the only thing to do. Herding our eight Creeps together, we guided them up the ramp and into the trailer where they would be safe while we were gone. Just as we had done this, Gerry, the missing snow goose, suddenly appeared through a line of trees bordering the field on the west side of the road a hundred yards away. She had walked all the way back to us! This seemed to impress the border officials even more than the fact that the flying geese had returned earlier. Knowing approximately where Middy had landed I spent a quarter of an hour searching for her near a cluster of farm buildings and was about to move on when she answered my calls. She had been hidden from sight in a depression in the ground, but now eagerly followed me back to rejoin the rest of the Creeps. One way and another the filming at the border post had been quite a nerve-racking experience for us, but for the border officials the free-flying geese had brightened up their day. So much so that the chief Customs officer asked if we could possibly send him a photograph of the geese at the border to illustrate the incident in their Customs magazine.

During May and early June we spent most of our time at Sand Lake for there was no need for us to reach the snow goose nesting colony at La Perouse Bay until it was time for the goslings to hatch around the third week in June. I phoned Dr Fred Cooke at Queen's University, Kingston, and he offered every assistance during our stay in the La Perouse Bay area. He and his associates have been studying the snow geese nesting there in much the same way as our friend Dr Charlie MacInnes and his team work at the McConnell River goose colony each summer. Fred Cooke mentioned on the phone that he and his students would be out in the colony every day during the hatching period and they would be happy to collect abandoned goslings for us.

Spring at Sand Lake was interesting for us, with much to film. We located several active dens of red foxes and spent many hours watching the antics of the tiny cubs as they played just outside the entrances, tumbling and growling until they were exhausted and collapsed in a heap to sleep. With the warmer weather the thirteen-lined ground squirrels popped their heads above ground and scurried around feeding on the new grasses after their long winter fast. Dozens of species of migrant birds arrived from their wintering grounds farther south. Some, like the snow geese, were merely *en route* to more northerly breeding grounds, while others nested on the refuge. We enjoyed the opportunity of photographing a wide variety of waterfowl, and became very interested in the work Lyle Schoonover was carrying out at Sand Lake with a breeding flock of captive giant Canada geese. The largest of the Canada geese, weighing twelve to fifteen pounds, this subspecies once nested widely over the northern plains but by the 1950's they were believed to be extinct. Then in 1962 some unusually large geese were noticed among a flock of Canadas in Minnesota, and plans were set in motion to reintroduce free-flying giant Canada geese to parts of their former nesting range.

At Sand Lake Refuge the first clutch of eggs is taken from the nest of each pair of captive geese and these eggs are hatched in incubators. The goose lays a second clutch about two weeks later and hatches these eggs herself. In this manner the maximum number of young are raised each year. Collecting the eggs is not as simple as it sounds, for the ganders in particular are extremely protective around the nest and can be most aggressive, hissing and striking at intruders with their powerful wings.

When fully feathered and old enough to fend for themselves, the young Canada geese are released in areas of the Dakotas where no waterfowl hunting is permitted, and it is hoped that there will soon be enough of them breeding in the wild for their future to be assured.

Among the many species of birds breeding at Sand Lake, our attention soon became riveted by the fantastic courtship behaviour of the western grebes. Some years earlier during a September visit to Bear River National Wildlife Refuge, in Utah, we had seen western grebes carrying their young on their backs as they swam, and had hoped one day to witness their amazing spring displays. Now we had the chance to do this in great detail.

By mid-May their penetratingly shrill whistles let us know that the western grebes had arrived on the refuge in force and had begun their courting rituals. We soon found that the biggest concentration was located near a small weir, on a seldom-visited part of the refuge, where between fifty and one hundred grebes could be seen swimming and calling at any hour of the day, immaculate in their black and white plumage with bright red eyes and needle-like yellow bill. Here and there a grebe dozed as it drifted on the surface, its long neck bent so as to rest along its back, but with its head always facing forward. Some fished, diving in the swift current just below the weir where a few cormorants and white pelicans often joined them. Occasionally a grebe surfaced with a small fish in its bill to present to a prospective mate as part of the courtship preliminaries.

Every now and then two birds would suddenly rear up out of the water side by side, and run across the surface in unison for up to twenty yards, with only their large webbed feet touching the water. During this display, bodies and necks are held vertically with heads at right angles pointing forward, and wings slightly open. At the end of the run both birds dive beneath the surface, to reappear a few moments later. Swimming together with necks stretching high, they move regally along, bending their heads every now and then to touch their own flanks in unison in a mutual display of mock preening. At first we were unable to anticipate the beginning of the grebes' paddling display, but after spending many days at the weir filming from a blind or from the Land-Rover, we learnt to recognize the more excited whistles that immediately preceded each run.

Exchanging these shrill whistles, two grebes swam towards each other with heads held low to the water. Maintaining this position the two paused about a yard apart, facing each other for up to a minute and taking it in turns to whistle and flick their beaks sideways across the surface of the water. They looked like two gladiators facing each other before engaging in deadly battle. Suddenly the pair rose simultaneously out of the water, and turning at right angles to face in the same direction, used their powerful webbed feet as sole means of locomotion as they raced across the surface of the water trailing a V-shaped wake. During the preliminary build-up a third grebe often swam over and aggressively broke up the show. Unfortunately we had no way of ascertaining the sex of any of the performers and were not certain that the two active ones were always a courting pair. If an intruder swam up just as the two grebes were about to rise up out of the water, all three birds sometimes participated in the running display.

In a two-hour period around the middle of the day on 22 May we witnessed a total of eighteen display runs by the western grebes, and on another occasion saw sixteen during a similar period in the late afternoon. Normally the displays occurred far less frequently than this, and during one five-hour wait we saw only four performances. At the end of May we noticed that a few of the grebes had brownish stains on their normally spotless white breasts and deduced that nest building was under way somewhere amongst the vast areas of cattails and phragmites nearby. Unfortunately we had to leave before locating any of their nests, but by then, the second week of June, their display runs were diminishing after more than a month of fascinating activity.

13 *Return to the Tundra*

We debated at length whether or not to take the Creeps and Fred back to the tundra with us, and finally decided that the extra work involved would be well worthwhile. Although our imprinted geese never seemed to resent being confined in a pen or in the trailer, we knew how much they would enjoy having the complete freedom of the tundra again. We also knew that we would miss them greatly if we left them behind, for they had been an important part of our daily lives for almost a year. Because of the shortage of space for travelling we reluctantly decided that Canada Goose would have to go to his new home at the Northern Prairie Wildlife Research Center, Jamestown, where we had left the pair of greater scaup the previous October. Here Canada Goose settled in well, enjoying the company of several other species of small geese – nenes from Hawaii, little Canadas from the Aleutians and brant from Alaska.

After obtaining the necessary permits and veterinary certificates for our feathered family we headed for northern Canada, reaching Churchill on 17 June. We soon had the Creeps installed in the disused Lambair hangar at the airport and ourselves in the Churchill Motel in town, for our camping gear was following on the next freight train. After the long train trip in their travelling boxes we let the Creeps fly each day, as we had no hunting problems around Churchill at this time of year. Fred made several flights with the geese, loudly and proudly using his new-found adult crane call and obviously very pleased to be back in the north. We were interested to see how much Fred enjoyed himself one particularly windy day at the airport when wind gusts of thirty-five miles an hour were recorded. Facing into wind and opening his wings he bounced into the air as high as ten feet at a time, a little like a child having fun on a trampoline. The Creeps were extremely cautious in this strong wind, preferring not to fly at all, but if they became airborne they were careful to keep their heads to wind as they tacked back and forth. But Fred had the time of his life, making several flights, manœuvring expertly in the wind and playfully chasing gulls that hovered overhead. He weighs about the same as one of the Creeps, but his wing surface is much greater, so that he has more control, and unlike the geese, can come in to land almost vertically, even in calm weather.

Although we had contacted our friend Bill Cooper, the Churchill manager of Lambair, in advance, schedules for charter planes in this part of the world are very uncertain. Operating in remote areas, with bad weather the rule rather than the exception, the pilots do an amazing job. All the Lambair planes were somewhere farther north when we reached Churchill and it seemed likely that we would have to wait several days before we could fly the thirty miles to the La Perouse Bay snow goose colony. The distance was short but there was no other practical way of getting there with all our equipment.

The morning after we arrived in Churchill we awoke to a world of white, a snowy surprise for us, and for the local residents as well. It continued snowing throughout that below-freezing day, depositing on Churchill 7.6 inches, an all time record for the month of June. Even during the whole of the previous winter the town had recorded a total of only 25 inches of snow. We were deeply concerned about the effect this storm might have on small nesting birds in the area and on the colony of snow and blue geese only thirty miles away, as the goslings were due to begin hatching any day now.

The snow certainly did not worry the Creeps; in fact they seemed to revel in it and had a great time bathing in the icy slush of partly frozen pools. Fred appeared to be worn out and we realized that he could not have slept at all during the train ride, for his eyelids kept closing as he stood alongside us. Finally he put his head along his back between his wings, tucked one leg up against his body, and slept soundly, something he only does when he feels completely secure. It is even more unusual to see him lie down to sleep, although he does this occasionally. In the wild, but never during the breeding season, large numbers of sandhill cranes usually roost together for the night in shallow lakes where they evidently feel safer than on land.

As nobody else had used the road before us, we had some difficulty driving through the deep snow on the side road to the hangar to let the Creeps out for exercise, but we finally made it. Later in the day we noticed a pick-up truck having difficulty on the same road, and before long the driver got out and walked across the snow towards us.

'Are you Mr Bartlett?' he asked. When I asked what I could do for him, he went on.

'My name is Wayne Collins and I would like to do an interview with you for C.B.C. radio about your snow goose filming. I was told to interview Fred too but when I asked for more information I was told to just come on out here and ask for Fred.'

'Well,' I said, 'Fred's right here, and I'm sure he'll be only too pleased to talk to you. But you'd better watch out that he doesn't peck holes in your microphone!'

When Wayne realized who Fred was we all had a good laugh. Strangely enough Fred seemed to suffer from stage fright and wouldn't even purr near the microphone, although previously we had never had trouble making sound recordings of him or of the Creeps.

During the afternoon of 20 June a Beaver aircraft returned to Churchill and was available to fly us the thirty miles to our camp site, but with our bulky goose boxes, plus camping and photographic equipment, it had to make four trips. Only three miles west of the snow goose colony there is a long dry esker where aircraft are able to land at any time during the summer, provided the wind is favourable, and this is where we had decided to camp. Nearby was an old army hut, large but leaky, with the impressive title of Camp Flicek painted over the door. Here we stored some of our supplies, for in late summer this region is notorious for polar bears and we had no wish to attract them to our tents by keeping food supplies anywhere near our sleeping bags! We carry no firearms, and until later in the summer when we were given a few thunderflashes (which resemble giant fire crackers), we had no protection at all against the bears except three healthy sets of lungs.

As soon as the plane had gone after delivering the last load of gear, we let the Creeps and Fred out of their travelling boxes. This was the first time we were able to have all of them free at one time since we left Sand Lake and they immediately took off and made a few quick circuits around the hut. Half an hour later the ten geese took off again to make a more thorough inspection of their new home territory. Far to the west they flew, until we could no longer see them with the naked eye, although from our slightly elevated position on the esker we had an uninterrupted view for miles in every direction. Watching through binoculars we saw with relief that at last they were beginning to fly a wide circuit of the area. Although they flew for a good quarter of an hour they were barely puffing when they landed near us. They certainly enjoyed flying more in the cooler northern air and acted like a bunch of happy school children on vacation. It was wonderful to have the Creeps completely free on the tundra again with no worries about hunters, or hazards such as tall trees, power lines and highways with speeding motor traffic. While the area where we camped is a good distance south of the tree line, there were no real trees within ten miles or more of camp, for the whole of the Cape Churchill area juts out into Hudson Bay and is covered with tundra-type vegetation. But the first thing we noticed was that the willow bushes here were much higher than at the McConnell River, sometimes reaching five or six feet, and occasionally on drier ground we came across an unhappy-looking, stunted conifer three or four feet high.

On our first day in the area, we decided to visit the biologists from Queen's University at their camp near the snow goose colony at La Perouse Bay, three miles to the east. This is a relatively new nesting colony, containing ten thousand adult snow and blue geese, which first began nesting here in small numbers in the 1950's. Donning hip boots and carrying back-packs loaded with cameras, we set out to hike across the boggy terrain. The Creeps and Fred elected to follow us, walking for much of the way, but flying to catch up whenever they lagged more than fifty yards behind. When we paused to rest and eat lunch beside a boulder on the watery tundra, the Creeps chewed on our pack straps for a few minutes before settling down to preen and rest. Fred wanted to try everything and we found to our surprise that he thought sardines eaten straight from the can made a tasty snack.

At the biologists' camp, which they had appropriately named Camp Quagmire, we talked at length with Dr Fred Cooke, Dr John Ryder and their students. Although the Queen's University team is mainly concerned with genetics, many other aspects of snow goose biology are covered by their studies. Since their arrival a month earlier, the students had worked in the colony each day mapping and numbering nests, noting clutch sizes and whether snows, blues, or a mixed pair belonged to each nest. Already they had collected four orphaned goslings for us and had them in a cardboard box near the heater in their tiny cabin. At Camp Quagmire everyone was fascinated by the Creeps and many photographs were taken as they bathed and swam in a stream running beside the camp.

'Aren't you afraid your geese will become affected by the wild geese?' Fred Cooke asked Jen.

'Not really,' she answered. 'We would be very surprised if they went off with the wild geese for good, but if they do it will be their choice; they are completely free day and night.'

As we approached Camp Quagmire on our hike that morning we had passed through part of the snow goose colony but the Creeps had ignored the wild geese completely.

That afternoon we were about to leave for our own camp when a large all-terrain vehicle suddenly appeared on the scene. Because of the direction of the wind nobody had heard its noisy approach until it was only a hundred yards from the camp. The biologists don't expect visitors out here, and having two lots in one day was unheard of! Two Royal Canadian Mounted Policemen clambered down from their iron steed and introduced themselves, saying that they were just checking to see who was in the area and making sure everyone's permits were in order. The Mounties kindly offered to give us a ride back to camp, so when they were ready to leave we gratefully tied our heavy packs securely on the back of the vehicle and climbed aboard. Whereupon the motor stubbornly refused to start and two embarrassed and frustrated Mounties spent the next few hours working as mechanics. Realizing that the delay might be a lengthy one, we reluctantly shouldered our packs and began the long trek back to Camp Flicek, taking it in turns to carry very carefully the cardboard box containing the first four orphaned goslings that the biologists had collected for us.

The Creeps were equally reluctant to leave Camp Quagmire, but finally we got them moving and they trailed behind us once more. Had we ridden on the all-terrain vehicle we are sure they would have been excited by it and flown all the way home without any trouble. As we trudged along we noticed that one of the Creeps, Whitey Two, stayed close to our heels and managed to keep up all the way back to camp without once attempting to fly. We thought that perhaps he had flown into willow bushes or hurt one wing in some way earlier in the day.

Whitey Two had never been a very fitting name, and now that he walked so well we renamed him Strider!

Fred and the other nine geese lagged behind from time to time and repeatedly made short flights to catch up. When we were about half-way home they flew on ahead to camp and we watched enviously. How we wished we could join them, for we hadn't yet got our 'tundra legs' and found the last mile or two very hard going. Slogging through the boggy patches our thoughts were: 'I can't rest here, it's too wet. I'll stop at the next dry spot.' Then on reaching a drier place the walking would be so much easier that we would think, 'I'll just try to keep going while I can. If I stop now it will be too hard to get started again.'

Tired though we were, our first concern on reaching camp was to care for the four tiny goslings which had become slightly chilled during the long walk. Placing them inside our shirts to warm up, we quickly heated water on our camp stove and filled a hot water bottle. Wrapped in towelling this was the best substitute we could offer for the warmth of an adult goose's downy underside. Exhausted, we had a quick supper and turned in for the night, placing the box containing the goslings alongside our sleeping bags so that we would hear their restless peepings if they became cold during the night. Sure enough they woke around 4 a.m. and we were very glad that we had taken the precaution of filling two Thermos bottles with hot water the previous evening and leaving them beside us. This year we did not have the luxury of the heated McConnell River cabin to ensure the goslings' well-being on cold nights and for the first two weeks we had to refill their hot water bottles once every night.

Soon after 5 a.m. next morning we heard the Creeps call as they flew off from camp while Fred and one goose answered from nearby. When we looked outside only Fred and Strider were in sight, and when the latter flapped his wings we realized why he was no longer flying. The annual moult of his flight feathers had begun. Already several primary feathers were missing from his left wing, and by evening all the longer main primaries had moulted from the tip of his right wing as well. From time to time Strider gave a few powerful flaps which often sent a few of his loosened feathers flying. After two or three hours the other nine Creeps announced their return by calling noisily as they came in for a landing at camp. We had no idea where they had been but assumed that they had found a good place to feed, for on subsequent mornings they flew away from camp very early, often accompanied by Fred. One morning, when the Creeps returned, ten geese landed near camp and came walking towards us. At first we thought we were seeing things, for Strider was still in camp. Then we realized the Creeps had brought back with them a single wild snow goose which flew off when the group came within fifteen yards of us. Poor Strider was a goose apart, staying close to camp and never letting us out of his sight if he could help it. Instinctively he seemed to know that we were his only protection, as we couldn't fly away and desert him!

'If the Creeps all stick to us as closely as Strider is doing, we'll have no worries about them leaving us when they are flightless,' I remarked to Jen and Les.

The Creeps did not know what to think of the new goslings and ignored them most of the time. Occasionally one acted aggressively, approaching with head held low and neck feathers quivering, but without ever harming the little birds. While our adult geese did not appear to pose any real threat to the welfare of our new charges, Fred was another story. To him the little goslings looked like an easy meal. At first we tried building a small Fred-proof pen so that the goslings could safely go outdoors whenever the weather was fine, but it seemed almost impossible to defeat Fred's hunting instincts completely. Even though the sides of the pen were made of plywood so that he could not see in, the sound of the goslings' peeping was enough to arouse his curiosity. Finally, after he learnt to dig underneath the boards at the side of the pen to peck at the goslings, we reluctantly decided that, until the goslings were a little bigger, Fred would have to be confined in an enclosure during the daytime. Each day the biologists rescued more stray goslings until we had fourteen altogether. Some were in very poor shape when rescued and we were greatly upset when we lost three of them.

After their early morning forays the Creeps spent most of their time around camp but continued to follow us whenever we went on filming trips. We enjoyed having them with us although we didn't always appreciate their efforts to help us. When we waded out into a large lake near camp to photograph common eiders, Arctic loons and herring gulls nesting on nearby islands, the geese loved to accompany us, swimming happily along behind in a tight group. Whenever we paused to set up a tripod in the shallow water they chewed at the feet and legs of our hip boots underwater, and believe me their powerful bills can hurt even through thick rubber. The legs of the tripod were an even greater attraction and it was impossible to take steady pictures when they began chewing on these. The female eider was incredibly unafraid, remaining on the nest when we were only a few feet away, with the Creeps making a lot of noise into the bargain. But the sitting herring gull barely waited for us to set foot in the lake before it left the nest to join its mate which was already diving and screaming furiously at us. Even though we were glad that the gulls never came as close as the Arctic terns had done the previous summer, the Creeps didn't like being dive-bombed and swam very close to us whenever the gulls were around.

By the end of June we noticed that the Creeps were no longer making such long flights over the tundra, although it was not until 1 July that any of the others became flightless. That morning the nine geese had made some short flights around camp, but in the evening when we went for a walk along the esker towards the coast, only seven could fly. Clark and Slacky – the two leaders – had now begun to moult their flight feathers and soon learned to walk beside us with Strider. The esker extended all the way to the coast, jutting out a short distance into Hudson Bay at Point Watson, about a mile north of camp. On pleasant evenings we found it most relaxing to walk out there after a long day of filming, especially as the ground was firm and dry and there was no need to wear waders. From the Point itself we had a clear view along the coastal flats where the snow geese now congregated with their goslings and at the beginning of July we saw our first polar bear of the season, sniffing its way along the high tide line in search of food.

A thick layer of white ice still covered most of Hudson Bay, but an ever-widening band of blue water had appeared along the coast and, after a north wind, large icebergs were left stranded on the tidal flats at low tide. Late in the afternoon of 3 July, after a busy day filming nesting birds far from camp, we noticed many more icebergs strewn along the shore than we had seen previously. Determined to photograph them at close range we shouldered our packs again and set out towards the coast, for the sun would not set until around 10.30 p.m. While we left the small goslings in their pen with plenty of food and water, Fred and the ten Creeps came with us. Once on the tidal flats we discovered that the stranded ice was much farther away than we had anticipated for these muddy flats may be as much as five miles wide at low tide. However, when we finally reached the icebergs there was no doubt in our minds that the hike had been worthwhile. The largest pieces were no more than six feet high and perhaps thirty feet wide, but they had been sculptured by the sea into a fantastic variety of beautiful shapes, tinted with blues and greens. In camp on calm nights we often heard the thud made by large chunks of ice breaking off these thawing bergs several miles away. We had hoped to film the wild geese near the ice formations but they moved out of camera range when we were still half a mile away. Fred and the Creeps had a great time puddling in the mud and pecking at the ice while we busied ourselves with cameras for an hour or two before starting the long walk back to camp. Now that the two leaders, Clark and Slacky, were flightless the other seven rarely took to the air.

The following morning we planned to hike two miles to a very boggy area to film whimbrel chicks hatching. Les had located the nest a few days earlier after a long search, and he checked the eggs each day so that we would not miss seeing the tiny chicks. Knowing that our route led through scrubby willow bushes for much of the way we felt that this would be a difficult trek for the flightless geese and tried to sneak away from camp without them seeing us leave. At first we thought we had succeeded but after we had gone half a mile we heard the Creeps calling as they flew towards us. As the seven fliers landed with us we wondered whether the three flightless ones had remained in camp. After waiting awhile to see if they appeared, we carried on for a short distance before pausing to photograph a stilt sandpiper on her nest, beautifully camouflaged in a tiny grass tussock. In fact we had only detected this nest a few days earlier by almost treading on it, which caused the bird to fly out from right under our boots. Just as we were ready to move on we looked back towards camp and saw our three flightless Creeps practically running across the tundra to catch up. As we carried on towards the whimbrel nest with all ten Creeps walking close behind, a lone snow goose veered from its flight path to circle several times overhead, descending to within thirty feet of us before deciding it did not really want to join the strange group below.

On reaching the swampy area where the whimbrel had its nest we found the chicks already dry and fluffy and roaming around the small mossy nesting hummock. Both parents were a short distance away calling agitatedly, but they soon settled down after we entered the blind and one returned to brood the chicks. By remaining with the whimbrels for several hours we were finally able to photograph from outside the blind without their being disturbed by our presence. In contrast to the exceptionally long curved bills of the parents, the two chicks had straight black bills only half an inch long, and distinctly bluish legs with very large feet. Their bodies were a light greyish-fawn with darker markings speckling their heads and backs.

All the time we were busy filming, the Creeps seemed happy just to be near us, alternately resting and feeding on the marshy vegetation. We appreciated their assistance as they thinned out the dense hordes of mosquitoes which settled on our clothing, and wondered how many of these tiny insects a five-pound goose would need to eat if no other food were available! Pommy was the champion mosquito catcher; we saw her catch as many as three mosquitoes with a single snap of her bill. The mossies really bothered the whimbrels, whose heads were constantly flicking to try to dislodge the pests. When we were ready to return to camp, we had a little trouble in convincing the Creeps to head in the right direction with us. It was a slow return trip with the three walking geese appearing to tire and frequently wanting to veer off course. Two days earlier the Creeps had followed Les to the same area when he took a blind out to the whimbrel nest, and on his own he had a lot of trouble in making them return to camp with him. On both occasions, as soon as camp came into sight the Creeps practically ran the rest of the way home. So we assumed that, through being unable to fly, they must have become temporarily disorientated amongst the high willow bushes.

Late that night as we walked to our tent after locking all our food in the hut, we noticed the Creeps and Fred feeding along the lake shore a hundred yards away. From past experience we knew that if the geese followed us to the tents when we turned in for the night, they would remain there pulling at the zippers and chewing on the guy ropes for an hour or more before becoming bored and moving away. So we tried a little deception, walking past the tents before doubling back to enter from the far side. Although this did not always work, Les told us later that on this particular night he looked out of his tent shortly after we had gone to bed and saw the Creeps standing in a puzzled group wondering where we had gone. Tired after our long day of hiking and filming we slept soundly until around 5 a.m. when we awoke to hear Fred and some of the Creeps near the tent. The geese jabbered away among themselves, while the rest of the Creeps seemed to be calling to them from farther away. Gradually the noises subsided and we dozed off again for half an hour. When we went outside the tent there was no sign of either Fred or the Creeps, but this was not at all unusual so we carried on with breakfast, caring for the goslings and doing other camp chores. At 8 a.m. we heard what sounded like two groups of Creeps calling to each other a few hundred yards north of camp. After walking a short distance towards the sounds and calling 'Creep, Creep – Creep!' once or twice, we remembered that we were about to do some filming with the small goslings and did not really want the Creeps back at camp helping with that project, so we called no more.

Two hours later a plane landed on the esker to pick up some of the biologists from Camp Quagmire and Fred flew back to camp alone to see what was going on. After reading a large batch of mail that had come in on the plane, we still saw no sign of the Creeps. Les walked down to Watson Point but without seeing or hearing them when he called. By lunchtime we were concerned, for the Creeps had never stayed away for such a long time. As we had last heard them just to the north of camp we felt that they might have wandered among the thick belt of willows towards the coast, and in their flightless condition could easily have got lost. Unable to fly and too short to see over the bushes, perhaps they could no longer tell in which direction camp lay. The previous evening, 4 July, we had noticed that two more of the Creeps had begun to moult their flight feathers, which meant that at least half of them were now flightless and we doubted if the remainder would attempt to fly on their own.

Unable to settle down to work, we began looking for the missing geese, although we felt fairly sure the Creeps would return to camp by nightfall. Splitting up so that we each searched a different area, we headed towards the coast, pausing every few minutes to shout: 'Creep, Creep – Creep!' We continued searching for the remainder of the afternoon and on returning to camp we compared notes. None of us had seen or heard any sign of the Creeps. That evening we were all hoarse from calling, but with the Creeps still missing we did not feel like talking anyway.

Lying in our sleeping bags that night we worried over the missing geese, knowing how exposed they would feel away from camp. Once we retired into the tents for the night poor Fred was completely on his own and seemed to miss the Creeps as much as we did. Later, I even found that I sadly missed the daily task of sweeping up their prolific droppings from the outer part of the tent and from the hut doorstep!

Throughout the following day we continued to search and call, with never an answering call from the Creeps, although once or twice wild geese along the coast seemed to respond as they sometimes will to a strange noise. Out on the coastal flats the wild geese moved off with their young when they saw us half a mile away. If we tried to get any closer they all swam out into the sea until we retired, for this watery retreat is their only means of defence against danger during the flight-less period. As we ate our supper dejectedly after the second fruitless day of searching, I remarked:

'I sure would like to know where the Creeps are just now.'

'I'm beginning to think they may have left because they wanted to go,' Jen said thoughtfully. 'It could be an instinctive thing with them at this time of year, during the moult, something stronger than the imprinting.'

'You could be right,' I answered. 'I've been wondering if their leaving could have something to do with the natural break-up of family ties.'

The one-year-old wild geese group together on the outskirts of the breeding colony when their parents begin nesting and raising a new family of goslings. The Creeps had stayed with us more than a month longer than wild snow geese their own age do with their real parents. As far as the Creeps were concerned, we were their parents and we had recently acquired a new batch of goslings. But the Creeps had shown neither affection nor dislike for the new goslings and in fact we doubted that the big geese even recognized them as young of their own species.

We had thought that we were doing the Creeps a favour by giving them complete freedom to come and go as they pleased. While they had the ability to fly this was probably true, but once they became flightless they needed our protection as never before, and in a way we had let them down. Each night when we retired into the tents to sleep we left them completely defenceless against roaming polar bears and wolves. In trying to reconstruct what had happened we soon came to believe that by leaving them outside at night we had unwittingly forced the Creeps to do what was best for themselves, and as a result they moved off to the comparative safety of the coastal flats where the wild geese also stayed. Normally geese rely on flying as their main means of escape from any danger, so it must be a traumatic experience for a young goose suddenly to find itself flightless for the first time since it was a gosling, and extremely vulnerable to a variety of dangers. Although at first we could not be certain that the Creeps were down at the coast among the wild geese, we soon had proof that this was so.

Fred was very lonely without the Creeps and followed us everywhere. Each evening when we went to bed he remained close to the tents picking mosquitoes off the outer walls, while we carried on a 'conversation' with him to let him know we were glad to have him near. Whenever we spoke his name or mimicked his purr, back would come his contented, deep purring reply. We didn't even scold him any more when he repeatedly pecked and pulled at our tent zippers as we did not want him leaving home too!

14 *Fly High and Free*

During the early morning hours of 8 July, while we were still half
asleep, Fred began calling as he flew off from camp towards the coast.
Shortly afterwards there seemed to be a lot of noise from some of the
geese down at the bay, but in our drowsy state we couldn't be certain,
and told ourselves that we must be dreaming if we thought Fred had
gone to visit the Creeps. However, at 4 a.m. on the following morning,
four days after the Creeps had disappeared, we instantly became wide-
awake when Fred began calling loudly just outside the tent. In the
calm early morning air we clearly heard geese answering him, and after
calling several times he took off and flew towards the coast, with the
goose calls becoming more and more excited as Fred's calls became more
distant. There was no mistaking the greeting he received when he
landed at the coast: the Creeps were very glad to see him! Although we
have several times seen Fred scare wild geese by flying near them, the
sounds we had just heard were definitely a greeting, and not alarm
calls.

Still not quite daring to believe what we had heard, we hurriedly
dressed. After smearing our faces and hands thoroughly with insect
repellent as protection against the hungry hordes of mosquitoes, we
set off for the coast about a mile distant. We reached the tide flats at
4.20 a.m. just as a huge red sun began to show itself, but a few minutes
later it disappeared behind a cloudbank and stayed there for most of
the morning. It was low tide and the wild geese were in scattered groups
a long way out near the water's edge. Scanning the area through
binoculars we had no luck in spotting Fred in the gloomy light; of

Jen and Des fit radio-controlled cameras under snow goose decoys

Des and Larry Leonard prepare the radio-controlled model plane for flight

We never tired of watching the 'Creeps' flying free at Sand Lake Refuge

The streamlined beauty of snow geese in flight

Two families of imprinted geese – a year's difference between them.
Three days after this picture was taken the 'Creeps' returned to the wild

In late July stranded ice floes litter the tidal flats of Hudson Bay

Scientists conduct banding drives when the snow geese and young are flightless

A snow goose and a blue goose mating – a rare sight in the nesting colony ▶

The goose incubates her eggs for just over three weeks while the gander stands guard

◄ Although onl a few hours o these goslings will shortly leave the nest

course we knew he would be much more difficult to see than the white geese among the large rocks. We began walking towards the geese, pausing every hundred yards or so to call loudly to the Creeps and Fred. Time and again we stopped, called, and moved on without hearing any response. Finally we had an answering call from Fred. We still could not see him but the sound definitely came from near the dark rocks towards the edge of the water where a group of snow geese were taking their young out into the sea, even though we were still a few hundred yards away from them.

Peering through binoculars we tried to find a group of ten geese with no young nearby, but it was an impossible task with several hundred adult geese in the water and along the rock-strewn shore. Fred certainly was not in the calm water, nor could we see him on shore even though we walked to within two hundred yards of the water's edge. He failed to respond to our repeated calls and we had almost given up all hope of locating him when I saw a crane settle on a grassy area of the coastal flats several hundred yards away. We thought that Fred must have taken off while we were concentrating on negotiating the slippery mud flats, and circled to land behind us. Relieved, we began to retrace our steps, but before we had gone fifty yards the real Fred surprised us by suddenly flying in from behind to land alongside. It was good to see him and there was now no doubt in our minds that he had been at the water's edge, and had flown down to the coast expressly to visit his friends the Creeps. He would not have ventured so far from camp completely on his own.

On our way back to camp we passed close to the spot where we had seen the wild crane land just before Fred flew to us, and found there were actually two cranes. Fred walked along just behind and as we approached the wild cranes they seemed unafraid and very interested in our little group. Soon we were less than fifty yards from the pair and we paused on a grassy mound to watch what they would do, as Fred walked calmly around us, gently picking the numerous mosquitoes off our trousers but completely ignoring the wild cranes. The pair began a cautious inspection, stalking slowly around us in ever tightening circles. On their second circuit they were only twenty-five yards away as we quietly sat down to rest our legs, whereupon Fred transferred his efforts to catching the mosquitoes massed on Jen's woollen cap but he still paid no attention at all to the wild cranes. On their third and final circuit the cranes were less than fifteen yards away. It was a fascinating and almost unbelievable experience to have this pair approach us so closely of their own free will, for in the wild, sandhill cranes are among the shyest and most suspicious of birds. Obviously their curiosity about Fred had overcome some of their natural fear of man, but we could not help wondering what they must have thought of Fred's un-crane-like behaviour!

Having completed their inspection the cranes moved a short distance away to feed and we slowly got to our feet and walked off towards camp. We had only gone a few yards when the wild cranes called. Each time Fred heard the sound he looked back briefly but made no attempt to answer. This meeting with the wild cranes strengthened our conviction that at this time Fred had no idea he was a crane. We felt that Fred, although raised with geese, long ago decided he was a human being, so strong was the imprinting where he was concerned.

Half-way back to camp Fred paused at a pond to feed while we carried on walking. When he flew to catch up, the wild cranes called to him from the coastal flats but Fred returned to camp with us. Later that morning Les saw the pair of wild cranes walk right past our tents and in the afternoon, when they called loudly from half a mile to the east of camp, Fred actually answered although he made no move to join them. Perhaps he was at last beginning to realize that he had something in common with cranes.

Originally we had planned to leave the Churchill area by late July, but with the disappearance of the Creeps we had to revise our plans. If we departed before they were flying again after their moult there would always be a nagging doubt in our minds that perhaps we had let them down, for we all felt that there was a slight chance that they might want to return to us once they regained their powers of flight. As the days and weeks went by we became more and more convinced that they would not return, that they had made their decision to stay with the wild geese, and the longer they remained with them the less likely they were to come back to us – or to any other humans. Nevertheless we altered our plans so that we could remain camped at Flicek until early August, when we figured all the Creeps would be flying again after the four-week flightless period. On most evenings we walked some distance towards the coast, often calling: 'Creep, Creep-Creep!', just to let them know we were still around.

Fred continued to visit the geese on the coastal flats in the early mornings, but as the days went by he flew down there less frequently.

Several times the wild cranes were near camp and gradually Fred became more friendly with them. One morning he was nowhere in sight when we came out of the tents and he failed to appear before we left on a filming trip. This was very unusual, for he normally flew into camp as soon as he saw that we were up and about. On returning to camp late in the afternoon there was still no sign of Fred and we became concerned. Setting off towards the coast we called loudly from time to time and were very relieved when he flew up from the coastal flats to land on the esker beside us.

The fast-growing new family of goslings had followed us from camp and before retracing our steps we paused to let them feed on the fine goose grass beside a shallow coastal pond. Fred began to wander off again and as we watched he joined the two wild cranes which had silently come to within thirty yards of us. There was little doubt that Fred had spent the whole day with his new-found friends. We followed as the three cranes walked up on to the esker, and after calling to Fred a few times we induced him to accompany us back to camp. Fred could now be free all the time, for once the goslings were two weeks old, he took little notice of them, and before long they were to take the place of the Creeps as his companions.

Two days after Fred's prolonged outing with the wild cranes, one of them paid him a visit in camp and gradually became his shadow. Although we have never known for sure whether Fred is in fact a male, we christened his new friend Lady. We have no idea what became of her wild mate for we never saw it again. Lady was soon an almost permanent fixture around camp, following Fred everywhere. She obviously wanted him as a constant companion, but during the daytime when we were about he usually ignored poor Lady. If she flew away from camp during the day Fred never followed her, but at night after we zipped ourselves into the tents we often heard them fly off together to feed. In the morning, as soon as he saw us moving about in camp, Fred flew back to greet us, with Lady sometimes following. Although we felt that Fred's ties to us were too strong for him to want to break them permanently, he too was free to return to the wild if he wished. Lady soon became amazingly tame, strutting between the tents and allowing us to walk within six feet of her. Her plumage was more rust-coloured than Fred's and her eyes a much brighter yellow, indicating that she was probably an older bird. Knowing that flocks of wild cranes are notorious for feeding in fields of ripened corn on their autumn migrations, we offered Lady some of the Creeps' left-over dry corn. She really enjoyed this and ate it from a dish beside the tent. Never having developed a taste for the dry kernels, Fred preferred his can of Miracle. Lady sometimes moved over to check what he was so keen on, but to her obvious bewilderment she could find nothing worth eating at all. When we threw a handful of raisins on to the ground for Fred, she would walk over to see what he was getting so excited about, but at first made no move to even pick any up, let alone eat them. However, before long Lady also acquired a taste for raisins and our supply dwindled rapidly.

During July, apart from filming the imprinted goslings growing up around camp, we found the nests of several other species of birds that we had not filmed the previous summer at the McConnell River. Altogether we located the nests of twenty-four different species of birds in the area, but the ones that interested us most were whistling swan, common eider, pintail, shoveller, least and stilt sandpipers, whimbrel, common snipe, horned lark and tree sparrow. Of these the whistling swan nest was perhaps the most interesting, for this region represents the extreme southern limit of the breeding range for this species. Conversely the shoveller at Cape Churchill is at the extreme northern limit of its breeding range and Fred Cooke told us that the two nests we found were the first ones discovered in the area.

Fortunately for us, a pair of whistling swans had made their nest on an island in a large lake only a quarter of a mile from camp. However this pair proved to be extremely shy, so at first we placed our photographic blind on another island almost a hundred yards from the swan nest. Over a three-week period we gradually moved it closer without disturbing them, until by the time the first cygnet hatched we were able to photograph the family scenes with the blind only fifteen yards away. With its silky white and grey down, and pink bill and legs, the cygnet was one of the most beautiful newly hatched birds we had seen. Within a few hours of hatching it was clambering about on its mother's back expertly catching mosquitoes. Some of our photographs have actually frozen numerous mosquitoes in mid-air as they buzzed around the swans, and we vividly remember the record number of mossies which plagued us in this blind.

The nest was a huge affair of grass and moss, almost two feet high and perhaps four feet across, with a deep crater in the centre for the eggs. At first there had been four large whitish eggs, but one disappeared without a trace of broken shell left behind. We noted that the pair took turns with the incubation for although the sexes are alike, one of our pair had more rust-coloured staining on its head and neck than its mate, and their bill markings were slightly different. Not long before the eggs hatched I was lucky enough to see the pair of swans mating in the water and was able to take some colour photographs. From these we found it was the male who did most of the incubating whenever we were in the blind, and the female seemed more timid. However, when the cygnets were hatching, the female was on the nest the whole time, with the male often sitting on the ground alongside the nest rather than floating on the lake as the off-duty adult had done during the long incubation period. From what we have since read about these swans, it is supposed that only the female actually incubates, but not very much detailed study has been done on their remote breeding grounds in the Arctic.

As the swans became agitated about the time the second cygnet emerged, we removed the blind so that they would stay at the nest until all the youngsters had hatched rather than swim off with only the first one or two. Unlike a goose egg, each swan egg gave us no warning as to when it was about to hatch. There was no pipping of the egg the previous day, but the cygnets appeared to hatch at one-day intervals. After the adults led their young into the water, the swan family remained on the nesting lake for only one day before moving quite long distances over-land to reach other lakes to feed and after a week we lost track of them.

During our first few weeks in camp we frequently heard the strange winnowing display flight of the common snipe. Flying so high as to be almost out of sight, the snipe dives steeply, causing its tail feathers to vibrate and make a strange whistling sound. With so many snipe in the area we kept on the lookout for their nests, but they are such secretive birds that we had no luck until we literally tripped over a sitting bird. We finally located three snipe nests with the sitting bird exploding from almost under our feet at the last possible moment to whirr away just above the marshy vegetation. Even when we knew the location of these nests their camouflage was almost unbelievable. Two were hidden beneath tiny willow bushes in boggy areas and the third was neatly concealed in a sedge tussock. We found this last nest on 4 July when it contained only three eggs, although by the end of the same day the fourth one had been laid to complete the clutch. Exactly three weeks later, on 25 July, the chicks hatched and left the nest, an event we un-fortunately missed due to dense fog.

Sitting in the blind photographing the snipe on the nest was a rather boring business as she often sat with her back towards us and we had to wait patiently for her to turn around. During these periods of waiting we were sometimes entertained by a small lemming that evidently lived underneath the blind. Every now and then it popped through a hole in the tent floor and scurried about near our feet looking for any crumbs of food that we might have dropped. This was the only rodent we encountered during our summer at Cape Churchill.

Whenever we walked out to the snipe blind, two miles from camp, we passed through what must have been the nesting territory of a pair of Hudsonian godwits. The pair always fluttered above our heads and tried to lead us away, then called noisily as they sat on bushes, where these long-legged waders looked strangely out of place. Although we searched at length we were never successful in finding the nest of this beautiful godwit, and wondered if perhaps the chicks had already hatched.

We had no more snowstorms, but the northern weather is never monotonous! A day with temperatures in the eighties could be followed by one with a maximum temperature of only 35° Fahrenheit. The ice did not completely disappear from Hudson Bay until the end of July, for the thaw was late this summer and we later heard that as a result of this many of the more northerly snow goose colonies had a poor nesting year. If the wind suddenly shifted to blow off the ice of Hudson Bay it brought about a sudden drop in the temperature and at such times many of the lakes near camp 'smoked' or 'steamed' and fog quickly formed.

The warmer weather was welcome although once again we did not appreciate the clouds of mosquitoes that quickly emerged. One female ptarmigan wandering with her young chicks had a mass of mossies like a dark cap on the top of her head and there was no way she could permanently dislodge them. The blackflies were very bad at times but we had learnt our lesson and used liberal applications of repellent to deter them from crawling inside our clothing. On calm nights the mosquitoes rained on our tents in unbelievable numbers, and at these times we felt so sorry for Fred that he sometimes slept in Les's tent to escape from the bothersome insects. We might add that it was never possible to house-train Fred and this tent will forever more carry the signs of his time in residence. The mossies mainly attacked the bare red caruncular area on the front of Fred's head, causing him to rub his head agitatedly along his back to dislodge the greedy bloodsuckers. They also settled on his long legs which we often smeared with insect repellent to give him some relief. Although he continued to enjoy eating mosquitoes, Fred discovered a new sport – catching the numerous large deer flies that settled on the outside of our tents from mid-July onwards. Sometimes he became so engrossed in this activity that he refused to leave camp to accompany us on walks.

Independently, all three of us had come to realize that Fred was a changed bird since returning to the north. This was not as a result of his association with the wild cranes, but because he seemed to have passed the childhood and adolescent stages of his life. During the first twelve months, much more individual attention had been lavished on Fred than on any of the Creeps, so perhaps it was no wonder that he had acted like a spoilt child much of the time. But now he seemed to have come of age. No longer did he peck at everything in sight just for the heck of it, he acted much more sedately in a manner befitting his regal appearance. He was like a puppy that had outgrown the slipper-chewing stage and had learnt to understand certain commands. Fred certainly knew by this time what we meant when we said 'No'. He knew the sound of his own name, and if we saw him start to chase the goslings and yelled, 'No, Fred!' he would suddenly stop and look away as if had been doing nothing wrong and was completely innocent. As soon as he thought our backs were turned, he would have another try at chasing them. But by the time our young geese were airborne Fred was on quite friendly terms with them, although in a strange way he appeared to resent their sudden ability to fly and sometimes flew after them, causing some to land. It was more of a game than anything else and most amusing to watch.

Early in the summer we had been worried lest Fred learn that a good meal of birds' eggs could be found in front of each of our little green blinds. Fortunately this did not happen; however he enjoyed playing with the small red marker flags that we placed a few yards from each well-camouflaged nest. Fred often pulled the flags up and had a great time throwing them into the air while flapping his wings and dancing up and down until he became bored and left them lying on the tundra, often far from the nest we thought we had marked so carefully! Although Fred helped us to lose the location of some nests he did help us to find others, especially willow ptarmigan nests. On two or three occasions when he was out walking with us, a male or female ptarmigan flew aggressively at him, letting us know there was a nest close at hand. Wild cranes are very partial to ptarmigan eggs and small chicks, and the ptarmigan did not want Fred anywhere near their nests.

Around the middle of July, when fledgeling sparrows and longspurs were leaving their nests, Fred learnt how to flush them out from their hiding places under tiny bushes. A few quick gulps and it was goodbye little bird. Although we hated to see Fred eat young birds, this only occurred on a few days during the peak hatching period, and he was after all only following his natural food-finding instincts. Short of keeping him shut up in a pen all the time there was little that we could do to stop him as he had more than enough food in camp. Another natural food that he was able to enjoy all summer was the small wood frogs that were quite common. He spent hours in and around the ponds near camp and we felt sure that he ate many other small aquatic creatures.

Every day now Fred used his new-found adult crane call, a truly wild sound that rolled out across the wide open spaces of the tundra. Sometimes it was in answer to the calls of the wild cranes, but more often he called when excited or alarmed about something. Once he began, Jen could easily encourage him to continue by mimicking his excited call. After one of these tundra sessions Fred prostrated himself on the ground just in front of Jen. Lying flat with his neck outstretched and the tip of his bill in the ground, he remained as if in a trance for perhaps half a minute before standing again. We also saw him do this three times in front of Lady. In later months he did the same thing several times with some of our new family of imprinted geese, and once even with a wild snow goose which he may have desired as a mate, for when she flew off he followed her for quite some distance before flying back to us. He also prostrated himself to two or three other people at odd times. Poor mixed-up Fred, he didn't know for sure if he was a crane, a goose or a human being. How we wished that we knew more about the intimate behaviour of the wild sandhill cranes so that we could better understand Fred's actions and know whether we had a male or a female crane in the family!

Lady often tried to entice Fred to dance with her by throwing pieces of moss into the air and dancing around with wings outstretched. If we were nearby Fred usually ignored Lady's efforts completely, but once, as we were returning to camp from far out on the tundra, we saw the two cranes clearly silhouetted on the ridge near camp. Both were leaping in the air and flapping, and their dancing continued for several minutes.

We were camped in one of the best areas in the world for seeing polar bears late in the summer. Fairly recently Canadian biologists have found a polar bear denning area in the forested area over a hundred miles south of Cape Churchill. In warm dens dug in the moss the female polar bear has her cubs in midwinter when snow blankets the ground. In late summer the bears move along the coast of Hudson Bay in search of food, for after the break-up of the sea ice in July they are no longer able to hunt the seals which, for most of the year, provide the major part of their food supply. Any of their droppings that we came across during July and August contained mainly grass fibres, and there were a lot of very hungry polar bears around. Once from an aeroplane at Cape Churchill we saw a bear carrying a snow goose in its jaws as it ran off across the tide flats, but there was no way of knowing if the bear caught a flightless bird, or whether the goose was sick or even dead when found by the bear.

Particularly in the autumn the polar bears tend to be a nuisance around the town of Churchill, and biologists have been trying to find ways of keeping the bears away from town. Many have been trapped and marked before being released as far from Churchill as the road system permits, which is not much more than thirty miles – a day's walk for a bear. More recently some of the nuisance bears were airlifted to restock a remote area several hundred miles away. In most parts of the Arctic and sub-Arctic around the globe polar bears have become scarce and Churchill must surely be the only place in the world that at certain times of the year can complain of having too many of these animals.

One day, soon after the ice had disappeared from the bay, we noticed what appeared to be a chunk of ice floating just offshore about two miles from camp. Checking through binoculars we saw that it was a polar bear swimming with her two cubs following in the water. They came ashore near Watson Point and spent three days within sight of camp. Walking part of the way to the coast, we sat down on the esker two hundred yards from the three bears to watch them as they moved about on the grassy flats. At first the female was fast asleep, sprawled almost flat with her cubs stretched out alongside her, and resembling some of the many light-coloured rocks in the area. Every now and then she lifted her head and used one huge front paw to wipe the annoying mosquitoes and blackflies from her face. Occasionally she raised her black nose high to sniff the air but seemed unaware of our presence. Finally she rose slowly to her feet, with the cubs following, and ambled about the flats in an apparently aimless fashion. After a few minutes she paused to roll over a rock with one front paw, then plonked her ample bottom into the depression where the rock had been. Here she sat and suckled her two cubs with her back towards us. Although we would have liked to take closer pictures of polar bear cubs, perhaps we were lucky that a female bear with cubs never came into camp.

Our main worry with so many bears around was for the safety of our new family of imprinted geese, for they only learnt to fly a few days before we left the tundra. Most nights they slept inside the old hut which had a very strong door, supposed to be bear-proof. The young geese would have made a tasty meal for a meat-hungry polar bear, and we were glad to have Fred around to give the alarm during the day, although at night he was usually away from camp with Lady. We joked that Fred was a human by day and a crane by night. Sometimes, soon after we came out of the tents in the mornings, we heard cranes calling and looked up to see Fred and Lady winging their way to camp together, but more often Fred flew in alone to say good morning to us, with Lady following some time later in the day. She seemed to us to be a mighty experienced seductress, and was always trying to lead Fred off. At times when she persistently tried to move closer to him in our presence, Fred actually seemed to be afraid of her and carefully kept one of us in between as a buffer.

During the latter part of July, while all the wild geese were flightless, we helped the Queen's University people with their goose banding. With the goose colony so close to Churchill they are able to use a helicopter, which is a tremendous help in locating and pushing together the goose flocks on the coastal flats. Whenever the chopper appears overhead the geese group together and people can be dropped off nearby to hold the geese or drive them towards a portable nylon-net pen. At the time of hatch, the students were extremely busy placing small numbered web tags on as many goslings as possible. By varying the position of these tiny tags on the feet of the goslings they can indicate the parentage of the youngsters – whether the parents are snows, blues or a mixed pair. During the banding drive permanent numbered leg bands are fitted and the complete information is recorded in a banding book. All the hard work by the biologists and students during the summer is, in a way, transferred to the numbered bands now carried by the geese for the rest of their lives. The parentage of snow-snow, blue-blue or snow-blue is recorded accurately, as well as the exact date of birth and the number of the nest (each nest being accurately plotted on a map) where it was hatched in the La Perouse Bay snow goose colony.

Once their banding operations were over late in July, all the biologists departed and we had the whole area to ourselves. Of course we had the polar bears for company, and there were lots of geese and other birds around, as their migration south would not be starting for another month. We thoroughly enjoyed the peace of those last two weeks, and it was somehow much more like the McConnell now with no planes flying around the area. We did not however complain when a chopper stopped unexpectedly one morning to drop off our mail on its way to Cape Churchill to help with a survey of Canada geese.

Whenever a caribou came anywhere near camp Fred's calls alerted us to its presence. He seemed to feel that large beasts had no right to be trespassing on his territory. Occasionally we saw two or three caribou in a group but there was one magnificent bull that stayed in the area for several weeks and frequently came inland to feed on willows near our camp.

We were occasionally able to obtain fresh meat when a plane came out from Churchill. One evening we had just finished the luxury of a steak dinner when Fred began calling loudly. Les looked out expecting to see a caribou, but instead found that a polar bear was watching the camp from only thirty yards away! The aroma from our steak must have been too much for the bear to resist, but as we moved about setting up tripods and cameras he waded out into a nearby lake. On hearing Fred's agitated calls Lady flew in to join the fun and the two cranes stood on the lake shore dancing up and down, flapping their wings, as they called loudly at the retreating bear. Climbing out on the far shore of the lake the bear gave a great shake, sending up a shower of spray from its shaggy yellowish-white fur. Unfortunately he did not go far before lying down among a small clump of willow bushes only three hundred yards from camp, with his nose raised sniffing the air. We felt sure he would come back during the night when all was quiet, but there was nothing we could do to prevent this.

Tired after a long day of filming, we decided to have an early night and had just climbed into our sleeping bags when two biologists, Bob

Montgomerie and Dave McIlveen, from the University of Guelph, arrived to spend the night in the old army hut. We called out to Bob and Dave from the tents, telling them to use anything of ours that they needed in the hut, and warned them that there was a polar bear in the vicinity. As one of their projects was to collect fresh polar bear scats for another scientist, this was cheering news to them!

'Don't worry about us,' Dave called back. 'We're just going to cook a couple of steaks outside, then we'll sleep in the hut. We have some special fireworks with us if the bear comes back.'

They walked on to the hut, and we had almost dropped off to sleep when Fred once again began calling loudly. We were certain this meant that the bear had been lured back by the smell of more steaks sizzling, but the window flaps of our tent were zippered shut from the outside and we were unable to see what was going on. Les could see out of his tent and gave us a running commentary. The two biologists did not know of Fred's reputation as a watchdog, and only saw the polar bear when it rounded the corner of the hut thirty feet from them. Shouting did little to deter the hungry bear, and even when a thunderflash went off with a very bright flash and a loud bang, the bear was reluctant to go. Using his shotgun, Dave fired a couple of bright flares which burst noisily near the bear and this finally persuaded it to leave.

'Hey, they're chasing it this way!' Les shouted.

Strangely enough, we could not get Les to answer any more of our questions and found out next day that he had gone to sleep. We were well aware that the determined bear would probably be back after things had quietened down, and we lay listening for a long time, but then dozed fitfully. Around 1 a.m. Jen was awakened by a scraping sound as the bear brushed against the canvas only a foot from our heads, trying to force his way into the front section of the tent.

'The bear's back!' she said urgently, bringing me instantly awake.

Sitting bolt upright, I gave a yell loud enough to wake the dead.

'Go on, get out of there!' I shouted at the top of my lungs.

There was no sudden reaction on the part of the bear, but he did turn around and moved slowly along the side of the tent inside the guy ropes, brushing against the canvas as he went, his massive silhouette clearly outlined against the fabric. As he reached the corner of the tent we both shouted again and the bear shuffled off. How far he went we had no way of knowing for the huge feet made no sound on the springy tundra. Apart from the scraping sounds as he rubbed against the canvas the only noise made by the bear had been its snuffly breathing, which we both heard clearly when it was very close. In lowered voices we discussed the bear and wondered how far it had gone from camp.

'At least I know it's possible to sweat on a freezing night,' I remarked to Jen.

'That's strange,' she answered, 'I have the shivers and can't get warm!'

It was a new experience for us, even though we often had elephants, lions, hippos and rhinos around our tents in Africa. For us the bear was much more frightening, as it was likely to break into our tent in search of something to eat. Many bears in the area acquire a taste for the foods associated with humans by visiting the garbage dump near Churchill. We did not have a gun, and there was absolutely nothing in the sleeping tent that could be used as a weapon. I remember having to resist the urge to get dressed – if the bear came back I would feel 'safer' facing it fully dressed. I can also recall thinking: 'This will be a marvellous adventure – if we live through the night!'

Next morning we were amazed to learn that Les had heard nothing at all of the bear's visit to our tent, although he was sleeping less than thirty feet away. And I'd thought that my shout was loud enough to wake the dead! On inspecting the main tent we found that, before it had woken us, the polar bear had evidently stood up and leaned heavily on one of the side flaps, ripping the stitching and leaving two big stretch marks on the canvas where its enormous paws had rested.

This incident occurred on 25 July and for the next three weeks, until we left the area, we saw an average of at least one polar bear a day. Before Bob and Dave returned to Churchill the following day they left three thunderflashes with us, so at least we could now make more noise if a polar bear came too close for comfort. Three loud bangs and we would be out of 'ammunition'! Fortunately most of the bears stayed down near the coast, but we had two more visitors around our tents at night. The next one was a smaller bear and was easily chased off when Les ran at it shouting. Unable to resist the temptation to try out his thunderflash, Les hurled it after the bear.

'I'm really sorry I threw the thunderflash,' Les said later. 'It seemed to scare the daylights out of the poor chap by the way it ran off.'

We wondered how many eighteen-year-old lads would be genuinely concerned about frightening a polar bear *too* much with a giant firecracker thrown from an otherwise unarmed camp! So we now had only two thunderflashes left. The third bear was larger and more persistent.

Strangely enough, as things turned out, I'd left Jen and Les alone in camp to face this bear. Because of our extended tundra stay I flew out to deal with business as well as to meet Julie, who was flying from Australia once again to join us for her school holidays. During the fifteen-minute flight to Churchill I counted eighteen polar bears from the air.

Later Jen and Les told us about their adventure with the third polar bear, which had arrived at the tents just after sunset. Fred and Lady ran towards the bear as soon as it appeared in camp. With their wings spread and uttering loud challenging calls, the two cranes moved to within twenty feet of the bear but it ignored them and carried on sniffing around the tent. Jen and Les also ran towards the bear shouting and beating on a large metal drum – it took little notice of them. After circling the tents several times, while expertly dodging the guy ropes, the bear stood with its huge front paws on the sloping roof of the main tent. Finding nothing of interest up there it dropped on to all four feet again to circle the tent once more. As Jen and Les could not drive the bear away they decided to sleep in the hut, saving the two remaining thunderflashes in case of further trouble. Next morning the tent was still standing, but with one side pushed in and a steel pole cracked from the weight of the bear. There were a few small rips in the canvas, none too serious, but it was certainly lucky for us that the tent had been empty when this bear came searching for food.

The Creeps were the only reason why we stayed so long on the tundra and the reader will be wondering what happened to them.

On 6 August we noticed that a few family groups of wild snow geese could fly again after the annual moult, and on succeeding days many more geese were flying along the coastal flats. We continued to walk down to the coast each day to look for polar bears and to let the Creeps know that we were still in the area. Early on the morning of the tenth many family groups began flying inland to feed around the lakes and ponds, some of them coming close to camp. Perhaps they were decoyed there by our young family of imprinted geese, which were now just learning to fly. Later that day there was an excited shout from Les:

'Where are the binoculars? Fred's out here flying with some snuggies!'

Rushing outside we saw Fred flying in close formation with five snow geese about three hundred yards south of camp.

'Come on Creeps,' we shouted in unison.

The little group banked and began to fly in a wide circuit around the camp, and although we continued to call to them they came no closer. After flying with the geese for almost a mile Fred finally responded to our calls and broke away to return to camp. The five snow geese carried on to land among the wild flocks at the coast.

There was absolutely no doubt in our minds that at least five of the Creeps were now flying; wild geese would never have allowed Fred to fly anywhere near them. But it was equally obvious that the Creeps no longer needed their foster parents. By their own choice they had made the transition to the wild flocks. All we could do was wish them well and hope that, guided by the actions of the other geese, they would continue to fly high and free. As they no longer wished to associate with us, we are now certain that they will never approach strange humans – and risk the danger this could bring them!

We had raised many wild creatures in the past, yet the Creeps had been magically different. Free to leave us at any time as we followed their wild relatives up and down the Central Flyway, they had remained an important part of our family for over twelve months while we travelled more than ten thousand miles together. Although our lives will never be quite the same now that they are gone, the Creeps taught us a great deal about the ways of snow geese, and we have many happy memories of shared experiences. Whenever we hear the exciting sound of wild geese flying overhead we always listen for the sound of the Creeps and find we can hear it in the call of each and every wild snow goose.

Epilogue:

What Happened to Fred?

Although the idea behind this book has been to tell the story of the snow geese, both wild and imprinted, somehow Fred, the sandhill crane, has made his presence felt between these pages and it would not be right to leave the reader wondering: 'What happened to Fred?'

We hated the thought of parting with him and kept him with us until he was more than two years old – and we ourselves moved on to Argentine Patagonia to film for eighteen months or two years. Perhaps we might have brought ourselves to part with him sooner, but on the journey south after our second tundra summer Fred suffered a near fatal illness while we were at Sand Lake Refuge. At this time Les had flown home to Australia but we still had with us our second family of young geese. We were due to fly to England to help with the final editing of the film at Survival Anglia Limited, London. The picture editing is one thing, but it was also important to go over the sound recordings with the sound effects editor since he had not personally heard any of the bird calls, nor the sound of a cannon net going off, and therefore really appreciated assistance in getting these things straight.

We returned to the refuge late one afternoon after a short trip to Aberdeen to send off the last of the snow goose film to London, and immediately walked over to the pen to let our birds out for a flight. There we found a very sick Fred. As Lyle Schoonover so aptly put it:

'Fred sure looks like he's got a whopping hangover, doesn't he?'

Poor Fred was unsteady on his feet and his head drooped, with nostrils watering and eyes closed much of the time. Although he ate a few grasshoppers when we offered them to him, he was not really interested and was unable to focus accurately on anything, at times missing the insects by an inch or more when he tried to take them from our fingers with his bill. Two of the young blue geese had developed similar symptoms several days earlier but were not nearly as badly affected and quickly recovered.

We decided that the only thing to do was to move Fred on to a large piece of canvas on the floor of the Bird House where we could watch him day and night. By nightfall he was very staggery and we wondered if he would last the night. Whenever one of us got out of bed to check he was either lying down or 'kneeling', and purred reassuringly in reply whenever we spoke to him. The following day he refused all food and even water and stayed lying down except that every hour or so he flapped and staggered to his feet so as to defecate without dirtying his feathers, and then flopped down again. The normally red skin on his head had faded to a dull pink and his eyes had a vacant, unseeing look, because the nictating membrane was not operating. We bathed his eyes every hour or two, but there was little else that we could do.

On the third day he drank some water and was on his feet a little, and we dared to hope that he might be really improving. But on the fourth day he was very low. Unable to stand, every hour or so he struggled convulsively to get up, and each time we would help by grasping each wing close to his body, then settle him down again when he was ready. Because his eyes were affected we kept Fred in a darkened room and tried to work on correspondence and sorting photographs in the next room. But with our spirits so low it was difficult to concentrate on anything, for it was as if we were helplessly watching a close friend or relative dying. Inevitably, in going through our photos we came across pictures of a healthy, active Fred and we could hardly bear to look at them. Every hour we were sure would be his last, and we realized that whatever his primary illness had been he now had such a serious respiratory infection that without drastic treatment there was little hope for him. But so little is known about the effects of drugs on many wild mammals and birds that while Fred had a chance of recovering on his own we felt that it was better not to seek professional help. However, I telephoned Johnny Lynch in Louisiana, knowing that in the past he had kept both sandhill and whooping cranes for the U.S. Fish and Wildlife

Service. Johnny told me that he had, in the past, picked up seriously ill sandhill cranes in the wild and injected them with antibiotics: some had died but some had lived, and so he recommended this 'shotgun' treatment for Fred.

We took Fred thirty miles to the Aberdeen Veterinary Hospital to see Dr Jim Prather, whose usual patients were farm animals and cats and dogs. In fact he had never treated a bird at the clinic until he had successfully pinned the broken leg of one of our young geese a week or so earlier. All of us felt it was almost a waste of time subjecting Fred to the injections, for he was now so low that we didn't see how he could possibly recover. All Jim Prather could do was to treat Fred with the same medication that he used successfully on cats and dogs with similar symptoms, and so he gave him two shots, one in each thigh. One was Vitalog – a cortisone – and the other an antibiotic called Daribiotic – a mixture of Neomycin and Polymixin B.

On the trip back to the refuge Fred lay quietly on the floor of the Chevy between Jen's feet, but at one point he suddenly gave a loud squawk and struggled violently as if having a convulsion, which ended with his head stretched out in front of him on the floor.

'I think he's gone,' Jen said sadly, as I pulled over to the side of the road and stopped. By bending low over Fred, we found we could detect very shallow and extremely rapid breathing. If before we had thought Fred's last hours had come, now we were sure it was only a matter of minutes. We sat there for a while with Jen holding Fred's head off the floor. Then, as there was no change, we drove slowly back to the refuge. We didn't speak but the same thoughts were going through both our minds – we would never again hear Fred's friendly purrs!

I carried him into the Bird House and placed him gently on the canvas, with his head propped between two plastic water bottles. He seemed to be in a coma and for two hours there was no movement, no sound, no change. Then his breathing gradually began to settle down to normal and later, when we held his head and dipped his bill into water, he drank a little. Soon he was making faint purring sounds again. But it was too slight an improvement to even begin to hope that he might pull through.

For four days and nights we had now watched over Fred, doing what little we could for him. Each night we were up every hour or so to help him whenever we heard him begin to struggle in the next room, and if we woke up and heard no sound we feared the worst. Sitting up in bed we'd call softly, 'Fred, Purrrp?', and be much relieved when back came his answering purr, though it was often very feeble.

But now we had another problem. On the day following our visit to Dr Prather with Fred we were due to fly to New York and later on to England, and a series of interviews and business appointments had been planned for us in advance. Although we had arranged to leave our birds in the big pen at Sand Lake where Lyle's daughter Sandra Schoonover would care for them in our absence, we couldn't expect anyone to cope with the twenty-four hour job of caring for such a sick Fred. And even if someone had been willing and able to do this we still could not have deserted Fred, for we were convinced that the only reason he had held on so long was because of his close attachment to us. With a marked lack of enthusiasm we both packed our bags that night, and when morning came and Fred was still alive, Jen remained behind to care for him.

Now that he was drinking a little water, Fred's only chance of recovery lay in getting food into him for the first time in four days; he was totally incapable of feeding himself. Apart from being unable to stand he had little control over his head and neck movements and could not cope with food placed in his bill anywhere near the tip. Therefore the only way to feed him was to kneel over him with one leg on either side of his body, open his bill slightly with one hand and with the other tip a small amount of Miracle from a teaspoon into the back of the bill so that he could easily swallow it. Fred soon became used to this system and by feeding him every hour or so he was taking a considerable amount of food each day. Jen's knees and back did not appreciate spending so much time down on the floor with him, particularly as her other main occupation was to catch two hundred or more grasshoppers a day to add to Fred's diet and this alone entailed considerable bending. Twice daily for four days Lyle helped Jen give Fred shots of antibiotic, and although he improved a little during this period he was still unable to stand or to feed himself. He now spent much time 'kneeling' although he first had to be helped into that position and frequently overbalanced. Each evening I phoned Jen from New York and anyone listening in on our conversation would have been somewhat mystified.

'How's Fred?' I would ask.

'About the same,' Jen replied. 'He's not really improving much but he is eating well. Today he had lots of Miracle, about two hundred grasshoppers, two frogs, and some raisins and green corn kernels.'

To try to get Fred using his legs Jen made a sling to support his weight. This was merely a rectangular piece of canvas with two holes for his legs, and the ends tied together above his back by ropes attached to a beam overhead. This worked extremely well and Fred spent more and more time in the sling with his feet just touching the floor and gradually began using his legs a little.

On fine days he now spent some time outdoors in the sunshine with the sling suspended beneath a large camera tripod, for his eyes at least were normal again and the bright light no longer worried him. What bothered the usually immaculate Fred most about his illness was being unable to keep clean and he lacked sufficient co-ordination to preen at all. After he had been ill for over a week Jen carried him to the nearby lake on a warm day and supported him in the sling while he stood in belly-deep water. At first Fred was a little scared but when he realized he was being firmly held he quite enjoyed it, repeatedly dipping his head underwater.

The most discouraging part of his illness at this stage was that although he was so much brighter in himself and no longer in imminent danger of dying, Fred was not improving as much as one would expect. Every few days Jen reported to Jim Prather by phone, but he could only suggest she begin giving him liquid vitamin drops while continuing to vary his diet as much as possible. When Fred had been ill for almost two weeks Jen took him in for Jim Prather to check over to see if he could understand why he wasn't improving more rapidly. By this time he could kneel for long periods and sometimes even stand for a few seconds before toppling over, but could not get into either position without help. There was no way that the vet could be sure whether Fred's lack of co-ordination was due to brain damage resulting from his original illness or whether it was a reaction to the antibiotics, so he didn't know if the condition might be temporary or permanent.

After almost two weeks of caring day and night for Fred, Jen was exhausted both physically and mentally. The staff of the Aberdeen Veterinary Hospital had become very interested in Fred, and, knowing that we should both be on our way to England, the two girls working there kindly offered to look after Fred while we were gone. He was now strong enough for someone else to take care of him without it affecting him adversely, and he could be left on his own at night. By day he could

spend most of the time outdoors in his sling under a tree and the only
real chore would be feeding him whenever one of the girls had a few
minutes to spare. If Fred had been showing steady improvement Jen
would have stayed until he was well enough to be left in the pen at
Sand Lake with the geese, but she could not go on looking after Fred
on her own without becoming ill herself. It was a difficult decision but
we knew he would be in very good hands. On the very day Jen left the
refuge to take Fred to the Veterinary Hospital on her way to the Aberdeen Airport, he remained standing on his own for half an hour in the
sunshine by the car while his bag was packed. Earlier that morning he
had, for the first time, raised himself unaided to the kneeling position.
This was the biggest improvement in his twelve days of illness and Jen
left him feeling somewhat cheered.

The next day we phoned the Veterinary Hospital from New York,
before flying on to London, and were relieved to learn that Fred had
settled down well. We had a busy time in England but often thought
about Fred and many people both there and in New York repeatedly
asked for news of him. Finally we received a wonderful letter from Jim
Prather:

'I am happy to inform you that Fred seems to have made a complete
recovery. He walks, eats and wants to fly or wander around the yard to
the point where we are going to send him back to the refuge. Have a
good time on your trip. Sincerely, Jim Prather.'

We felt like rushing out of Anglia's London offices and shouting
the good news to everyone passing by in Park Lane, but fortunately restrained ourselves or we might have wound up in a mental
hospital.

Weeks later we returned to Sand Lake and Fred seemed as pleased to
see us as we were to see him. Lyle Schoonover's young daughter Sandra
had taken very good care of him and the geese during our absence and
we learnt that when Fred was out of the pen he often walked around
the Bird House looking for us. He didn't seem to have got all his old
zip back and we felt we couldn't leave him to face a cold northern
winter, and so we took him south with us to Arizona together with
four of the geese to keep him company. One of them, Slob, had earlier
broken a leg when making an unwise downwind landing on the lake,
narrowly missing the bow of our canoe but smashing heavily into a
paddle. Jim Prather had operated on the leg to pin it, and this had
healed beautifully although Slob still had a slight limp. We felt that he

would also benefit from a warmer climate but we could not cope with taking all the geese as we would be very busy preparing for a prolonged expedition to South America.

We drove to Jamestown to leave the other seven young geese at the Northern Prairie Wildlife Research Center and visited Canada Goose whom we hadn't seen for six months. Although he showed no outward sign of recognition or affection, we noticed that while the other geese near him went on feeding or preening while we were there, Canada Goose never took his eyes off us. There was no doubt that he remembered.

During the following months we saw a great deal of Fred and the four geese, although we had to leave them several times in the care of Jay Helm while we flew to the Bahamas to visit our base and do some writing, and to Australia to see Julie and our relatives. After completing *The Incredible Flight of the Snow Geese*, it took many weeks to clean and repair all the equipment in preparation for shipping it to Argentina for our next major filming project (Dr Roger Payne's study of the southern right whales). While there was a chance to spend some time with Fred we didn't want to part with him until we were finally leaving North America.

Fred was now just like a dog, following us everywhere outdoors and wanting to be inside with us whenever possible. He was quite well-behaved indoors now and he sometimes stood for hours in the middle of the floor preening, leaving behind lots of little feathers and flaky bits of scale, as well as the inevitable dropping or two. If we were typing he couldn't resist pecking at the moving keys, but his favourite toy was a typewriter eraser in the shape of a pencil with a little brush on one end. He would take this off the table and play with it on the floor for a long time, repeatedly stabbing at the brush with his pointed bill and purring excitedly. Sometimes he became quite interested in a television programme and on more than one occasion jumped up on to the table for a closer look at the screen. Whenever he heard the click of the refrigerator door being opened or closed Fred became very excited because he knew only too well that the cheese and cooked chicken he loved so much were kept inside.

In late March, Fred seemed very animated and called frequently. Several times a day he flapped and danced, jumping and using his bill to pick up and throw twigs or large feathers into the air. Perhaps his built-in time clock was telling him that this was the season to begin the journey north to the tundra, for at this time the wild lesser sandhill cranes that winter in New Mexico, the Texas panhandle and Mexico, are beginning their flight to northern nesting grounds. During their spring migration as many as 200,000 cranes may gather along a 150-mile stretch of the Platte River in Nebraska, and during the latter part of April they pass over the Dakotas on their northbound flight, usually flying at much higher altitudes than snow geese. Because of depredations by cranes on cereal crops, some provinces in Canada and states in the United States have opened hunting seasons for cranes in autumn. While geese, and especially ducks, can withstand a certain amount of hunting because of their high rate of productivity, it is doubtful if cranes can do so because, although they lay two eggs, they rarely raise more than one young to maturity each year.

Fred now learnt to make a strange new sound. Whenever he was particularly annoyed he lowered his head until it was almost upside-down between his legs and, simultaneously fluffing up his feathers, he emitted a rumbling growl from somewhere deep inside. The whole performance lasted for several seconds, after which he carried on normally. We usually put him in the trailer just as darkness fell but some evenings we left him out until a little later. Then he would patiently stand on the porch beneath the outside light, watching us at work through the glass doors. One evening he was standing there where we could see him clearly, when suddenly there was a rustling of feathers and Fred had disappeared. Calling loudly to him, we searched the neighbourhood frantically with torches for half an hour until finally we found him limping along a track towards us. Evidently he had been frightened either by a dog or a coyote and had taken to the air to escape. But once airborne he could not see well enough to make a proper landing in the darkness and must have crashed into a bush, for one leg was cut and bleeding. He must also have bruised himself internally, for we had a worrying three days as he refused to eat. Then he ate another of Miss Harrington's goldfish for his first meal, but she readily forgave him. He soon recovered, but we decided that he was not equipped for night flying and after this incident made sure he was safely in the trailer before darkness fell.

From our second family of young geese we learnt a great deal and were able to make some interesting comparisons between them and the Creeps. When the second group were about five months old they were much more advanced in plumage development than the Creeps had been at the same age. At first we could not believe this, but we had many dated photographs taken of the Creeps at various ages and there was no doubt that our new snow geese were whiter in November than the Creeps had been in January. Trying to deduce the reason for this we figured that it must have something to do with the second group being subjected to far less stress than the Creeps, for they were rarely confined in travelling boxes and were allowed much more complete freedom, as we reached Sand Lake several weeks ahead of the hunting season. This also meant that group two were feeding on green corn as part of their diet almost two months earlier than the Creeps and this also could perhaps have had something to do with their precocious development. Maybe being in a warmer climate at a younger age could have helped too. But proof that stress played a major part in the difference in plumage development was provided by the fact that Slob, the snow goose that broke his leg, lagged far behind the rest of the snow geese in this respect. He was only off his food for forty-eight hours, but was definitely upset by being kept separated from the other geese and being unable to use his leg or fly for two or three weeks while convalescing.

In mid-June when they were almost a year old, we noticed that our four geese of group two suddenly began making agitated nasal talking noises. They would keep this up for perhaps an hour at a time, two or three times a day, and this behaviour was always accompanied by repeated flicking of their heads as if they were trying to dislodge invisible insects. Perhaps some inner stirrings were telling them it was time for something to happen, though they knew not what; after two or three weeks this activity ceased. We are fairly certain that the Creeps made similar sounds just before they left us for good, and were only sorry that we weren't able to be on the tundra again with our new geese in order to learn more from their behaviour. Another comparison we were able to make just before parting with the four geese concerned the moulting of their flight feathers. This did not begin until they were thirteen and a half months old, a month later than is normal for the wild geese on the tundra. The delay was probably due to their remaining in a warm southern climate instead of migrating north where the daylight is much longer each day, but the moult was so well synchronized that all four became flightless within three days.

At the end of the first year Fred apparently had not moulted his flight feathers and he certainly did not lose the ability to fly. However when he was about twenty-four months old he surprised us by suddenly losing most of his primary and secondary wing feathers and was grounded for over three weeks. Strangely enough on each wing he lost only the innermost of the five main primaries. The remaining four old primaries remained in place until Fred's other new wing feathers had grown and he was flying again. Then we noticed new main primaries growing among the old ones near his wingtips, and presumably the old feathers fell away later after we finally parted with Fred.

Not trusting any airline to care for the unaccompanied birds when they finally travelled to North Dakota, I made detailed arrangements to fly on the same plane with the geese and Fred before leaving for South America. The same Frontier Airline plane flew from Tucson to Bismarck, with stops at Phoenix, Denver and Rapid City and I had a window seat where I could watch to make sure the birds were not mistakenly off-loaded at any of these stops. In a paper bag beneath my seat I carried Fred's can of Miracle and corn for the geese in case 'we' were off-loaded along the way. Fortunately this did not happen and Lyle Schoonover met me at midnight when the plane touched down at Bismarck. It was great to see him again and we had plenty to talk about while waiting for Fred and the geese to be unloaded. But they were not on the plane! After frantic phone calls we tracked them down – still sitting in the freight section at the Tucson Airport! Needless to say I was extremely concerned for the birds' welfare and very unhappy with the airline. I talked to the man on duty in Tucson and he promised to feed and water the birds for me. They finally arrived in Bismarck after a twenty-four hour delay. We loaded them into Lyle's station-wagon for the one-hundred-mile drive to Jamestown in the early hours of the morning and Forrest Lee, the biologist in charge of the birds at the Research Center, waited up to receive them at 2 a.m. Every comfort had been provided for their needs, including a large indoor pool where they could swim and clean up after the long journey, and there were even grapes for Fred!

It is now August, 1974. We are camped on Whale Bay, Valdez Peninsula, Argentina, and this is where the final pages of this book have been written. Needless to say, we often think of the Creeps, Fred – wishing he was here catching grasshoppers with us – and our second group of geese. Those at Jamestown are doing well and Lyle Schoonover tells us that Fred is as tame as ever and has a handsome set of new grey feathers. He spends his days out in the grounds catching worms and insects, but goes into the Hatchery Building each night. There are other cranes at the Research Center and we hope that, given time, Fred will come to believe he is a crane after all and find a mate among his new friends.

He is happy and fit at Jamestown, North Dakota, and is very well cared for by the men of the U.S. Fish and Wildlife Research Center. They enjoy having Fred around!

Acknowledgements

The experiences recounted in this book could not have taken place without the co-operation, help and shared enthusiam of a great many people. Many of them are mentioned in the text but we owe an extra 'thank you' to the following.

We are especially grateful to Aubrey Buxton and his team at Survival Anglia Limited for financing the long and costly snow goose filming project and for having faith that a worthwhile TV Special would result. Colin Willock, particularly, was enthusiastic about Snow Geese from the very beginning and much of the success of the finished film was due to his instinctive feel for the subject. Although this is not a book about the film itself, we appreciated Les Parry's inspired editing of the film (along a story-line provided by Colin) for which he received both an Emmy and an Eddie Award in the United States: a unique honour for an English editor!

Advance planning was tremendously important to the success of this filming project, and we would like to thank Graham Cooch, Sam Jorgensen, Charlie MacInnes and Paul Prevett for their help and sound advice. Without the co-operation of officials of both the Canadian Wildlife Service and the U.S. Fish and Wildlife Service in obtaining the necessary permits, we could never have raised the Creeps nor travelled with them as part of our family. Government and private veterinarians aided us in Canada and the United States, but we particularly appreciated the friendly help of Dr Jim Prather and his co-workers at the Aberdeen Veterinary Hospital.

It would be impossible to name all the individuals who assisted us in the field, but we want to thank collectively the many employees of the U.S. Fish and Wildlife Service who, without exception, helped us in every way possible. Often they acted as if *we* were doing *them* a favour in making a detailed film on the snow geese! At Sand Lake Refuge, where we spent the most time of all, Lyle Schoonover aided us above and beyond the call of duty (we hope his family has forgiven us for all the free time he spent with us), and it is impossible to thank him adequately for so generously sharing his time and knowledge. Our thanks also to Don Snider for, among other things, introducing us to home-smoked chicken. On our visits to Squaw Creek Refuge, Harold Burgess was always enthusiastic with his help, and we thank the Burgess family for their wonderful hospitality.

At the McConnell River we were particularly grateful to Charlie MacInnes and his students from the University of Western Ontario for their cheerful comradeship and support; Charlie took us 'under his wing' from the very beginning. During our second summer on the tundra we greatly appreciated the co-operation of Fred Cooke and his team from Queen's University. While we were in the Churchill area Bill Cooper and Father Paradis individually helped us in many ways, typical of people living in the Far North where lives depend so much on others.

Our deepest thanks go to our nephew Les Bartlett for his enthusiastic help throughout the eighteen months he spent with us on Snow Geese, and to Lee Lyon for her cheerful contribution during the early months of the project. They both helped tremendously with the hard work of looking after the big family of geese.

The fact that we finally completed this book is partly due to the enthusiasm, and gentle prodding, of Billy Collins, who with his wife Pierre was able to spend a few happy days with our 'family' and, as usual, Fred completely stole their hearts. We have greatly appreciated the editorial guidance given us by Marjorie Villiers, Ernestine Novak and Adrian House. Our thanks also to the *National Geographic Magazine* for allowing us to use their excellent snow-goose-migration map in this book.

For finding time in their busy lives to read and comment on the manuscript we are extremely grateful to snow goose experts Harold Burgess, Graham Cooch, Fred Cooke, Charlie MacInnes and Lyle Schoonover, but we accept full responsibility for the views expressed in this book. Our thanks also to Mel Wuersig, the first person to read the manuscript when it was still 'hot' off the typewriter.

We particularly want to thank the Slats Helm family and Emily Harrington for being such good friends to Fred and the Creeps. As visitors to the United States and Canada we have been impressed from the very beginning by the warm-hearted helpfulness of everyone we met in both countries. So finally we would like to thank the many 'unnamed' kind and considerate people we encountered while following the 'Flight of the Snow Geese': friendly people in immigration, customs, post offices, motels, garages, on planes . . . everywhere. Our thanks for making us feel completely at home.